5th EUROPEAN CONFERENCE ON EDUCATION

10 – 13 SEPTEMBER, 2000

AT

**SEALE-HAYNE FACULTY
UNIVERSITY OF PLYMOUTH
NEWTON ABBOT
DEVON
TQ12 6NQ
UNITED KINGDOM**

"FROM PRODUCTION AGRICULTURE TO RURAL DEVELOPMENT : CHALLENGES FOR HIGHER EDUCATION IN THE NEW MILLENIUM"

Conference Programme,
Text of Keynote Papers,
Abstracts of Contributed Papers
and List of Participants.

Edited by Eirene Williams

SPONSORS

The International Conference Committee and the Local Organising Committee are extremely grateful to the sponsors listed for their generous support for various aspects and activities of this Conference.

The World Bank
Food and Agriculture Organisation (FAO) of the United Nations
South West of England Regional Development Agency
The Smallpeice Trust
Blackwell Science
Teignbridge District Council
Plymouth City Council
Taste of the West
The Dartmouth Smokehouse
Rocombe Farm Fresh Ice Cream Ltd
Gray's Farm Cider
Westmoor Quality Foods
Peninsula Milk Processors Ltd
Yarner Spring Water
Western Morning News

Published by the University of Plymouth, January 2001

Seale-Hayne Faculty
University of Plymouth
Newton Abbot
Devon TQ12 6NQ
UK

tel: +44 (0) 1626 325800
fax: +44 (0) 1626 325657

ISBN Number 1-84102-074-5

ENGINEERING YOUR FUTURE

The Smallpeice Trust is an educational charity
founded in 1966 which aims to:
- enhance perceptions of engineering as a career
- offer young people an awareness of the career opportunities available within professional engineering
- provide basic and support skills as well as encouraging personal development all of which are immediately useful in an engineering career
- create worthwhile links between education and industry

The Trust provides an ongoing programme of subsidised training courses, one of which is the Engineering Careers Foundation Year (ECFY). This is a "gap year" programme aimed at students with deferred entry to university on an engineering-related degree.

What does the course provide?

ECFY is designed to provide post "A" level students with a "gap year" to remember. The course concentrates on providing students with the basics of engineering and associated skills which will make them more immediately useful and hence employable at the end of their degree. It is not designed to replace all or any part of university education but to complement it. The course is divided into 3 parts.
1. Academic study
2. Foreign language tuition
3. European work placement

The 12 week academic study covers such subjects as basic engineering skills, supervisory management and computer-aided design (CAD). In the second part students develop their foreign language skills on a 4 week intensive course at a European language school. Finally students undertake a 13 week placement in one of 9 European countries where they actively contribute to the company and work on projects as they arise.

The complete course is largely funded by the Smallpeice Trust, requiring students to contribute only a small amount per week towards subsistence.

Who will benefit from the course?

Usually we find that ECFY students: 
- want engineering experience before university but would like to go abroad for a few months
- want to meet new people and enjoy the social scene
- want to learn about management techniques, interpersonal skills and teambuilding in preparation for a future career
- know that Europe is getting "smaller" and want to develop a foreign language as well as experience European work methods and culture.

For further information contact:

The Smallpeice Trust
Holly House, 74 Upper Holly Walk, Leamington Spa CV32 4JL
Tel: 01926 333200 Fax: 01926 333202
Email: gen@smallpeicetrust.org.uk Internet: http://www.smallpeicetrust.org.uk

International Conference Committee prior to ECHAE5

Chairman:
Dr. Robert van Haarlem, The Netherlands
Wageningen University, P.O.Box 910, 6700 HB, Wageningen, The Netherlands.

Committee:
Dr. Eirene Williams, United Kingdom.
Seale-Hayne Faculty, University of Plymouth, Newton Abbot, Devon TQ12 6NQ, United Kingdom

Dr. Jozsef Kiss, Hungary.
Szent Istvan University, Pater K.U.1, Gödöllö, Hungary

Dr. Elmer Stuhler, Germany.
Munich University, Hochaker Str. 7, D85350 Freising, Germany

Dr. Bill Lindley, formerly FAO, Italy.
Viale delle Terme di Caracalla, 00100 Rome, Italy

Dr. Vasili Lavrovskij, Russia.
Timiryazev Agricultural Academy in Moscow, ul.Timiryazevskaya 49, Moscow, Russia

Local Organising Committee at Seale-Hayne Faculty, University of Plymouth.

V. Crean
Dr. J.C. Eddison
S.J. Fisher
Prof. F. Harper
J. Palmer
R.J. Soffe
M.A.Stone
M.F.Warren
Dr. E.N.D. Williams

THE CONFERENCE

The University of Plymouth was pleased to host this 5th European Conference on Higher Agricultural Education at the Seale-Hayne Faculty of Agriculture, Food and Land Use, located at its Newton Abbot campus in the county of Devon in the south west of England. Previous conferences in this successful series have been held in The Netherlands, Hungary, France and Russia. The conference provided the opportunity to hear high quality presentations and to engage in discussions with participants from a wide range of countries on a topical subject of special interest to those who are active in Higher Agricultural Education and related subjects.

THE TOPIC

From production agriculture to rural development : challenges for higher education in the new millennium. It was clear from the outcome of the fourth conference in Moscow that this is a topic of growing importance to educators, extension workers and policy makers in all European and many other countries. The issue of rural development is compelling and the higher education system needs to react to the significant and rapid changes that are occurring in rural areas and communities. This conference addressed the implication, for centres of higher education, of the changing focus from production agriculture to rural development and posed the following questions for discussion:

- **What are the implications for institutions that have long specialised in higher agricultural education?**

- **How are universities and other centres of higher education adapting to the changes?**

- **Is there a danger of neglecting the fundamentals of agricultural education?**

- **What developments are required in the curriculum and the delivery of learning during transition?**

THE OVERALL OBJECTIVES OF THE CONFERENCES

- to contribute to the global agenda for Higher Agricultural Education in the third millennium.

- to formulate strategies for greater international understanding.

- to enhance co-operation through networking.

- to share experiences and outcomes from the best current national and international projects to improve the overall quality and relevance of the teaching and learning process.

CONTENTS   Page

Keynote papers

1) RURAL DEVELOPMENT: THE ISSUES   13
Chair : **Clare Broom**, Dean of Seale-Hayne Faculty, University of Plymouth, United Kingdom.

Professor R.J.Bull, Vice-Chancellor, University of Plymouth, United   15
Kingdom.
Welcoming Address and opening paper on The University of Plymouth and its role in a rural region.

Professor P. Lowe, Centre for Rural Economy, University of Newcastle-upon-   19
Tyne, United Kingdom.
The challenges for rural development in Europe.

Prof. Dr. Ir. Guido van Huylenbroeck, Department of Agricultural   32
Economics, Ghent University, Belgium
Rural development and the environment : how to remunerate farmers for their contribution?

Professor A. Pilichowski, University of Lodz, Poland.   40
Rural areas in transition : the sociological perspective.

Professor A. Errington, University of Plymouth, United Kingdom.   56
From agriculture to rural development in the curriculum and in the rural economy.

2) FROM 'AGRICULTURAL' TO 'RURAL' : THE CHALLENGES   69
FOR HIGHER EDUCATION.
Chair : **Dr.Vasili Lavrovskij**, Timiryazev Agricultural Academy, Russia.

Dr. J. Rowinski, Institute of Agricultural Economy, Polish Academy of   71
Sciences, Poland.
Challenges for higher education in Poland in the new millennium.

C. Maguire, Senior Development Specialist, The World Bank, USA.   87
From agriculture to rural development : critical choices for agricultural education.

L. Gasperini, Senior Officer, Agricultural Education, FAO, Italy.   105
From agricultural education to education for rural development and food security.

Prof.Dr. C.M. Karssen, Rector Magnificus, Wageningen University and   113
Research Centre, The Netherlands and President of the Inter-University Conference for Agriculture and Related Sciences (ICA).
Implementing change : challenges for institutional structures and management.

Contributed papers

1) AGRICULTURAL EDUCATION IN TRANSITION – FACING UP TO THE ISSUES.   119

Chair: **Dr David Gibbon**, Freelance consultant in Agriculture and Rural Development, United Kingdom.

P.Brassley, University of Plymouth, United Kingdom.   121
Education and professionalisation in English agriculture.

M.Lostak, Czech University of Agriculture in Prague, Czech Republic.   122
To survive the change or to cope with it? (Education for Rural Development).

P.Stonehouse, University of Guelph, Canada.   123
An inter-disciplinary approach to addressing sustainability issues in the agri-food sector.

A.Kancs, Institute of Agricultural Development, CEE, Germany.   124
Integrating CEE countries into the EU : a challenge for rural development research.

K.Balazs and L.Podmaniczky, Institute of Environmental Management, Hungary.   126
Rural enterprise planning : an applied approach to address new challenges.

T. Dunning and E.Hesselink, Larenstein International Agriculture College, The Netherlands.   128
The philosophy and principle behind the curriculum in a new course in Rural Development and Innovation.

2) AGRICULTURAL EDUCATION IN TRANSITION - BENEFITS AND EXPERIENCES FROM CASE STUDIES.   129

Chair: **Prof. Jozsef Kiss**, Szent Istvan University, Hungary.

P. Busmanis, Latvia University of Agriculture, Latvia.   131
Higher agricultural education and research in Latvia – results of reforms and future challenges.

S.Figiel, S. Pilarski and Z.Warzocha, Olsztyn University of Agriculture and Technology, Poland.   133
From agricultural to agribusiness and rural management education : a case study from Poland's transitional economy and university environment.

Y.Bicoku, S.Androulidakis and J.Phelan, Institutional Support Unit, International Fertiliser Development Center (IFDC), Albania.   135

Strengthening the co-operation between higher education and rural economy : continuing education and integrated extension curriculum development in two Albanian universities.

L.Bartova*, L.Tauer** and A.Bandlero** Slovak Agricultural University, Nitra, Slovakia.** Cornell University, USA.
The Institute of Economic Studies and transformation of economic education at the Slovak Agricultural University in Nitra.
137

M.J.Linington and F.M.Ferriera, Vista University, South Africa.
Refocussing higher agricultural education to address rural development in South Africa.
138

V.I.Georgievsky , A.Ivanov, N.S.Sheveljev and G.D.Afanasjev, Timiryazev Agricultural Academy in Moscow, Russia.
Farm animal behaviour and zoo-psychology of domesticated animals – a new course for Russian Agrarian Universities.
139

3) FACING CHANGE IN THE CURRICULUM AND ITS DELIVERY.
141

Chair : **Prof.Dr. Jozsef Kiss**, Szent Istvan University, Hungary.

M.Warren, University of Plymouth, United Kingdom.
Farm Management : the death of a discipline?
143

T.Wieczorek, Warsaw Agricultural University, Poland.
Developmental tendencies of agricultural education in Poland.
144

A.Cooney, University of Saskatchewan, Canada
Agribusiness Management Development Program : stimulating rural development by facilitating the entrepreneurial spirit in Saskatchewan.
146

J.Driesse, van Hall Institute, The Netherlands.
Educational improvements and innovation in life sciences and nature conservation. Key factors for safe food in a sustainable world.
148

M.Slavik and I.Miller, Czech University of Agriculture, Prague, Czech Republic.
Developing the pedagogical competency of a university lecturer.
150

A.Varea, University of Cambridge, United Kingdom.
Agricultural education in transition : the role of educational research.
152

4) CURRICULUM CHANGES AT THE INSTITUTIONAL AND REGIONAL level.
153

Chair : **Prof. Peter Stonehouse**, University of Guelph, Canada.

V.Chumakov* and T.Bruening**. *Moscow State Agroengineering University, Russia, ** The Pennsylvania State University, USA. 155
Industrial development through international programs at Moscow State Agroengineering University.

G.Fuleky, Szent Istvan University, Hungary. 157
University integration : the most up-to-date challenge for agricultural education.

K.Parton and T.Funk, University of Guelph, Canada. 158
Evaluating the Guelph Electronic MBA in Agriculture : next step globalisation.

O.Meixner, University of Agricultural Sciences, Austria. 159
Economics of mountain systems – a network of European universities training regional specialists for the European mountain areas.

A.Popov, T.Korsunova and V.Radnatarov, Buryat State Agricultural Academy, Russia. 161
The experience of the Buryat State Agricultural Academy in improving ecological education in training students of agriculture.

G.Skrzypczak and S.Drzymala, Agricultural University of Poznan, Poland. 163
Agricultural University of Poznan curriculum changes in the last decade and new ideas for higher education in future.

5) AGRICULTURAL TO RURAL DEVELOPMENT EDUCATION IN PRACTICE. 165

Chair : **Prof. Dr. Pavel Kovar**, Vice-Rector International Relations, Czech University of Agriculture, Prague, Czech Republic.

E.Ramos and M.Delgado, University of Cordoba, Spain. 167
From reflection to action : Higher Education in rural development at the University of Cordoba.

D.Gibbon, and J.Jiggins, Swedish University of Agricultural Sciences, Sweden. 168
Internationalisation, interdisciplinarity and learning systems for transforming agricultural education.

A.Koutsouris, Agricultural University of Athens, Greece. 169
Sustainable rural development and curricula in higher educational institutes : issues and challenges.

D.Hyman and C.Holsing, The Pennsylvania State University, USA. 171
Graduate education via the Internet : a world campus graduate certificate in community and economic development.

B.Rivza* and M.Kruzmetra.* Chairperson of the Council of Higher   172
Education, Latvia.
Sustainability of rural areas in transition and higher agricultural education.

6) EDUCATIONAL INNOVATIONS WORKSHOP.   175

Chair : **Dr.Wiebe Nijlunsing**, van Hall Institute, The Netherlands.

J. Polson, Ohio State University, USA.   177
Producing quality video for use as a teaching aid.

V.Mothes, Institute of Agricultural Development in CEEs (IAMO), Germany.   179
Model based on online agricultural education in central and eastern Europe.

T.Felton and M.Stone, University of Plymouth, United Kingdom.   181
The interactive footpath : an on-line tool for land-based professionals.

M.Stone and N.Witt, University of Plymouth, United Kingdom.   183
The development and use of a managed electronic learning environment for
agricultural education and training.

M.David, University of Plymouth, United Kingdom.   185
Overcoming marginality : cyber training in cider country.

M.Stone, R.Williams, R.Soffe and S.Fisher, University of Plymouth, United
Kingdom.   187
The theoretical underpinnings of student centred learning. A case study based
on business management teaching and learning for agriculture and related
programmes of study at the Seale-Hayne Faculty, University of Plymouth.

M.Stone, R.Williams, R.Soffe and S.Fisher, University of Plymouth, United   189
Kingdom.
Student centred learning in practice. A case study based on business
management teaching and learning for agriculture and related programmes of
study at the Seale-Hayne Faculty, University of Plymouth.

M.Stone, R.Williams, R.Soffe and S.Fisher, University of Plymouth, United   191
Kingdom.
Developing student centred learning in the information age. A case study based
on business management teaching and learning for agriculture and related
programmes of study at the Seale-Hayne Faculty, University of Plymouth.

M.Stone, R.Williams, R.Soffe and S.Fisher, University of Plymouth, United   193
Kingdom.
Student attitudes to Student Centred Learning. A case study based on business
management teaching and learning for agriculture and related programmes of
study at the Seale-Hayne Faculty, University of Plymouth.

POSTER PAPERS — 195

A.N.Kartashevich. Belarusian Agricultural Academy, Belarus. — 197
Adapting the curriculum of higher agricultural education in transition

R.Y.Kasimov, I.Gotovtseva and R.G.Akhmetov. Moscow Agricultural Academy, Russia. — 198
Qualimetric model of assessment of results of higher agricultural education.

T.M.Korsunova, A.P.Popov and M.B Tumanova. Buryat State Agricultural Academy, Russia. — 199
The agrarian production ecologization in the Baikal region as the basis for transition to sustainable development.

P.Kovar. Czech University of Agriculture in Prague, Czech Republic — 201
International Hydrology Course – an example of a water-orientated postgraduate study programme at CUA Prague.

V.Linnik*, E .Korobova *, A.Kuvylin *, M.van der Perk and P.Burrough**.** — 202
* Russian Academy of Sciences, Moscow, Russia, ** Utrecht University, The Netherlands
GIS modelling for training in decision making on safe agricultural production in contaminatedareas (Novozybkov case study).

A.N.Novikov and B.Y.Korniyenko. National Technical University of Ukraine, Ukraine. — 203
Optimal control of the granulating process of fertiliser in the fluidised bed.

V.Shepovalov. Timiryazev Agricultural Academy in Moscow, Russia. — 204
Agricultural education and reformation technological reconstruction of plant growing.

S.Tzortzios and G.Adam. University of Thessalia, Greece. — 205
Adapting Computer-Based Learning Methods in Agricultural Education

M.Warren. University of Plymouth, United Kingdom. — 207
Farmer use of the Internet and the implications for technology supported distance learning.

I. Yamada. Ministry of Agriculture, Forestry and Fisheries of Japan, Japan. — 208
Farmers' opinions about programs of agricultural experience for children.

CONCLUDING PAPER — 209

Prof. Fred Harper, Emeritus Professor of Agriculture, University of Plymouth, United Kingdom — 211
Summary presentation reporting the main outcomes of the Conference.

LIST OF PARTICIPANTS

Participants are listed
1) by country in alphabetical order of country, and
2) by family name alphabetically.
Addresses and other contact details are given

Rural Development: the issues

Keynote papers

The University of Plymouth and its role in a rural region

John Bull

Vice-Chancellor, University of Plymouth, Drake Circus, Plymouth, Devon, PL4 8AA, UK

1. May I welcome you all most warmly to the UK, to the county of Devon, and of course to the Seale-Hayne Faculty of the University of Plymouth. The fact that this is the fifth European conference since 1992 on higher agricultural education indicates not only the importance of this topic, but that the issues facing production agriculture and, more broadly, rural development remain high on the strategic agenda not only for Europe but also for much of the developed and developing world.

2. For Universities in Western Europe who have and do provide courses and research in agriculture, the shift in demand from production agriculture to broadly defined rural development - diversification into tourism, heritage, alternative land use, and small businesses (increasingly 'e' based) - poses considerable strategic and policy issues. Poor and declining agricultural incomes, poor images (of genetic engineering, of animal welfare, of a variety of food 'scares') and, from a UK perspective, an unhelpful EU agricultural policy - all combine to make the undergraduate and postgraduate study of agriculture and its upstream derivatives less and less popular.

3. Yet it is vitally important that at least some Universities sustain an active and substantive presence in the broadly defined agricultural disciplines to provide a source of new knowledge, to provide the skilled workforce and to ensure that there remains an independent and critical questioner of both government policy and commercial practices in food and food related industries.

The University of Plymouth

4. The University of Plymouth clearly identifies with and struggles with these issues. One of the largest of the UK Universities - some 25,000 students (around 2000 of whom are from 100 countries abroad), 3000 staff across a complete range of disciplines (shortly, incidentally, to include undergraduate medicine) - we have as part of our mission, a commitment to support the regional agenda, to assist in this region's economic, social and cultural well-being.

5. We describe ourselves as a 'distributed' University, delivering courses and research at four main locations in the south west peninsula, with our campuses scattered along a 100km chain - Seale-Hayne almost in the middle, with Plymouth to the west and our Exmouth and Exeter campuses to the east.

6. That distribution gives us a critical presence in both the urban and rural locations, underlining our commitment to both. Before turning in a moment to share with you some of the ways in which we try to respond to our region, particularly the more rural part of it, let me remind you briefly of some of the characteristics of this part of the UK.

The South West of England

7. The South West of England is a region which stretches from Bristol in the north, to Bournemouth in the east to Penzance in the west, and nearly 40% of its 4.6 million population live in areas with a population of less than 10,000, the highest proportion of any of the UK regions.

8. The South West shares the generally distinctive features of all rural areas:

    i) Firstly, there is a prevalence of small firms operating often as a family business. In the counties of Devon and Cornwall some 70% of registered companies employ 5 people or less.

    ii) Secondly, there is a wide geographical dispersion of businesses and households, and the modern developments in communication technology reduce but do not eliminate the innate disadvantages of distance which give rise to higher production and distribution costs for many rural businesses, and make household access to goods and services difficult and expensive. The threat, and for some, the reality of social and economic exclusion is ever with us.

    iii) Thirdly, the South West has a wide variety of public goods - the national parks, the flora and fauna, the landscapes - which are highly valued and which will require a trade off of rural regeneration schemes which will raise incomes but not damage the public goods.

9. The South West shows all those generic features of a rural region, but there are also some distinctive features:

    i) In terms of annual gross output, the agricultural sector of the South West is bigger (at c£2.8 billion) than in any other UK region, and in many of our sub regions over 10% of the workforce is engaged in farming or growing.

    ii) Only one UK region attracts more UK tourists than the South West, and some 17 million tourists spending around £3billion come here each year.

    iii) Nearly 40% of the land area of the South West is designated as an area of outstanding natural beauty contributing both to local quality of life and, of course, to tourism.

    iv) Fourthly, a glance at any map of the UK will show that the South West has by far the longest coastline (by a factor of 2) of any UK region, resulting in many coastal settlements highly dependent on the sea and coastline for income generation.

    v) Finally, whilst most UK regions with large rural areas also have large metropolitan cities, the South West has very few. In Devon and Cornwall, the largest city, Plymouth, has a population of only 250,000 - leaving the remaining 1.75 million largely living in small towns and villages.

The University and its Region

10. So how does this University try to respond to a region which has both generic and distinctive rural characteristics? How does this University seek to ensure that its central purposes - teaching and research - are made accessible to such a community?

11. Firstly, I offer both a paradox and a tension. We can only help this region in the long run if we are or become world class. It is our national and international reputation which will attract outstanding staff, students, research, and thus offer real benefits to the community, and with scarce resources, that means we must sometimes say 'no' to requests for local involvement in the short term, to short lived schemes which have palliative but no sustained benefit.

12. That said, let me offer some specific illustrations of the practical ways in which we seek to respond:

    i) **Partnerships with Further Education (FE) Colleges** We fund and quality assure the provision of some 4000 full and part time places on mainly sub degree (Higher National Diploma) course in 16 FE Colleges in the region, offering local provision particularly for mature students for whom the time and cost of travelling to a University campus is prohibitive. Many of those students transfer with enhanced status to the final years of our degree programmes.

    ii) **High Speed Networks** We lead a South West HE consortium which links all those institutions, together with the FE colleges to a high speed, wide band network (nicknamed SWAN) which itself is linked to the UK, HE computing network, offering access to teaching and research materials and to the web. Our technological expertise offers substantial support and benefit to small HE institutions and the FE Colleges. We are currently exploring links to the growing schools networks.

    iii) **Strategic Collaborations** We work closely with other Universities in the region to attract funding which alone we could not achieve. For example, our collaboration with Exeter University has enabled us together to obtain government funding for the first new UK Medical School for 40 years, offering a real prospect of improved health care within the South West peninsula - provision of advanced medical and surgical procedures for which people currently have to travel to other parts of the UK.

    iv) **Rural Area Training** We lead a consortium which has set up, with the aid of EU funding, some 40 'drop in' networked centres in rural areas of Devon and Cornwall, through which businesses and individuals can obtain information, advice and training tailored to their needs when they need it.

    v) **Consultancy** We provide a range of consultancy through a variety of mechanisms to local businesses. For example, the Teaching Company Scheme, funded in part by Government, enables a business, particularly small businesses, to identify a major problem it wants tackled, links a key employee with a University mentor, and over a one or two year period finds solutions - and often enables the employee to obtain a masters degree. We currently have 28 such schemes. Through our Food Technology Centre we work closely with our Ministry of Agriculture, Fisheries and Food to provide a range of advice, consultancy and short courses for local businesses.

vi) **Inward Investment** The University works closely with government and local authority agencies engaged in attracting inward investment to the region. We are - or rather our research is - often one of the critical factors in persuading businesses to come to the region.

vii) **Applied Research** Much of our research is deliberately at the applied (as opposed to the generic) end of the research spectrum, and we attract some £4 million of funding from industry, some local. We 'spin off' technology and science based companies into a 'Science Park' jointly owned by ourselves and the City of Plymouth.

Agricultural Education

13. I could continue at length - you will be pleased to know I will not! - but I hope I have illustrated how this University plays its part in supporting what is essentially a rural region. Your conference is addressing one, but a very key, aspect of that interface, the particular dynamics of the production agriculture and rural development, and our Seale-Hayne Faculty hosting this conference has much to offer to that debate.

14. Working with declining industries is never easy; persuading them to transform and transmute is difficult. But Universities cannot and must not walk away from the problems by disinvesting themselves. We have a moral and intellectual responsibility to play our part in assisting rural economies to achieve both social and economic regeneration. I hope that the conference in enabling you to share the insights and activities of colleagues from Europe and beyond will contribute to that end.

The Challenges for Rural Development in Europe

Philip Lowe

Centre for Rural Economy, University of Newcastle upon Tyne, NE1 7RU, UK

1. Introduction: Rural Europe at the Millennium

The term 'Rural Europe' conveys a misleading notion of unity. The characteristics that are common to the rural areas of Europe far outnumber the differences. Rural areas in European countries are characterised by relatively low population densities, compared to urban areas, and by relatively extensive land uses - such as agriculture and forestry. Yet beyond these two simple characteristics, Europe's rural areas are extremely diverse in their socio-economic conditions and natural and geographical circumstances and, therefore, in the nature of the development prospects and problems they face.

Because of this diversity, it is difficult to point to a single, over-arching 'rural problem' to be addressed through rural development policy. In the past, the broad parameters of what constituted 'rural development' were easier to agree than is the case today. This was partly because they were defined in terms of the common external goal of 'modernising' agriculture and rural services, in order to catch up with the standards of urban sectors and areas. Not only has the centrality of agriculture as the defining feature of rural areas diminished but there has also been increasing recognition in recent years that 'development' is not a uni-linear process, but is much more complex and multi-dimensional. It is accepted too that the development of rural areas should much more build upon and conserve their intrinsic qualities and assets. Thus what may be required in one rural locality may differ sharply from prevailing needs and opportunities in another.

One common factor facing all rural areas, however, is the process of reform in the public policies addressing rural development. For example, the 1990s saw a series of steps to reform rural and agricultural policies in Europe, including changes to the Common Agricultural Policy (CAP) and the Structural Funds of the European Union (EU). Further reforms are highly likely in the coming years, particularly given pressures from world trade negotiations and from the closer relationship between Central and Eastern European countries (CEECs) and the EU. In the CEECs, new rural development and agricultural policies are also being adopted in anticipation of accession to the EU. Thus while land uses and socio-economic conditions vary greatly, rural areas are bound by a collective need to respond to a public policy framework that has become increasingly Europeanised over the past decade.

Three main sets of concerns inform these policy reforms. *First* is a desire to equip economic sectors and individual firms with the capacity to adapt to the market conditions of an increasingly liberalised system of world trade. *Second* is that economic development should be sustainable, especially in environmental terms. *Third* is that development policies should be flexible and more locally tailored to meet the diverse needs and conditions in rural areas. Successive waves of CAP reform, and the evolution of the Structural Funds, have brought an increasingly territorial approach to rural development, and the establishment of rural development as the new 'second pillar of the CAP', and this is reflected in the pre-accession SAPARD programmes being promoted with the CEECs. However, serious concerns remain about the limited scope of policy reform and the continued difficulties in resolving the various economic, social and environmental problems experienced in Europe's rural areas.

In this paper I want to explore the diverse perspectives on rural development found across Europe and to consider some of the implications of the Europeanisation of rural policy.

2. Deconstructing the Rural Development Agenda

Not only do the basic social conditions, land uses, and economic structures vary greatly across Europe's rural areas, but so also does the notion of 'rural development'. This can make discussing rural development difficult, because actors from different countries may have quite contrasting ideas about what constitutes the legitimate scope of rural development policy. It is therefore useful to be clear about the nature of the contemporary rural development agenda in Europe, the way that this agenda has evolved in recent years, and the diversity of national and institutional perspectives that inform contemporary rural development.

The Exogenous Model of Rural Development

The classical formulation of the rural development problem, which was dominant in post-war Europe, was founded in an understanding of urbanisation and industrialisation as mutually reinforcing and unilinear processes whereby capital and labour were increasingly concentrated in cities. Within the modernist development trajectory, the function of rural areas, stripped of other economic activities, was to provide food for the expanding cities. The notion of balanced or articulated development was embodied in the achievement of a spatially polarised but nationally integrated geography in which cities, functioning at the core of specialised regional economies, concentrated the bulk of population and commercial and industrial activity, while rural areas became dominated by a technically progressive, market-orientated agriculture, pursued by full-time, professional farmers. The spatial category of rural was often viewed as a residual category and became equated with the sectoral/occupational category of agriculture.

The 'problem' of rural development followed from this classification and was seen to arise in those regions and countries where too many people remained on the land, thus restricting the transfer of profit and labour needed to fuel urban and industrial growth, as well as inhibiting the development of a competitive and efficient agriculture. It was widely believed that such stagnant regions needed to be connected to dynamic centres and expanding sectors. It was never clear, however, what the eventual equilibrium between urban centres and regions and their rural hinterlands would be. Even areas of highly commercialised agriculture seemed destined to steadily lose population because of the tendency towards diminishing returns within agriculture. Thus even the most developed and prosperous rural areas were locked into an unequal exchange relationship with urban-industrial growth poles.

Classically, therefore, the development problems of rural areas and regions were diagnosed as those of marginality. As a concept, marginality has a number of dimensions - economic, social, cultural and political - although in discussions about rural development marginality is often understood in geographical terms to be synonymous with peripherality or remoteness. In this sense it has long been recognised that people living in rural areas have suffered problems of physical exclusion from urban-based services and jobs. Low productivity in the primary sector has compounded such difficulties, condemning those who live and work in rural areas to a low standard of living.

Peripherality, though, was always a metaphor for other types of distance too. Rural areas were distant technically, socio-economically and culturally from the main (urban) centres of activity. In all of these respects they were backward. While steps could be taken to encourage the transfer of progressive technologies and practices from dynamic sectors and regions, it was only through overcoming peripherality that rural 'back-waters' could be reconnected to the main currents of economic and social modernisation. Within this fundamentally exogenous perspective on rural development, the basic policy response was a combination of subsidising the improvement of agricultural production to enhance farm incomes, and the encouragement of labour and capital mobility (see Table 1).

Table 1 Exogenous Model of Rural Development

> **Key principle** - economies of scale and concentration
> **Dynamic force** - urban growth poles
> The main forces of development conceived as emanating from outside rural areas
> **Function of rural areas** - food and other primary production for the expanding urban economy
> **Major rural development problems** - low productivity and peripherality
> **Focus of rural development**
> - agricultural industrialisation and specialisation;
> - encouragement of labour and capital mobility

The state-sponsored modernisation of rural services and of agricultural practices and technologies has been a constant feature of post-war rural development. Policies to encourage labour and capital mobility, though, have fluctuated (Clout, 1993). The first phase in European policy was one of consolidating farm structures (i.e. land reform in southern Italy and Greece, and plot consolidation and enlargement programmes in France, West Germany, Spain, and the Netherlands) linked to land improvement schemes (including drainage and irrigation) and the development of farm-orientated infrastructure. The aim was to establish commercial units able to mechanise and absorb other 'productivist' technologies and to reduce the agrarian population particularly through the elimination of small and marginal holdings.

However, it became apparent that such measures could not stabilise rural economies and rural populations; indeed, they seemed to intensify the flow of labour out of agriculture and often out of the rural areas altogether, promoting concern for the viability of certain regions. The depopulation of peripheral areas and the balanced development of national territory became preoccupations of policy in countries such as France, Sweden, Austria and Switzerland. A second phase of rural development, typically focussing on peripheral regions, therefore, emphasised the attraction of new types of employment into rural areas. Processing facilities were established and manufacturing firms were encouraged to relocate from urban areas or to set up branch plants. As well as financial and fiscal inducements, development agencies concentrated on providing infrastructural support, including improvements in transportation and communication links, power supplies and the provision of serviced factory sites and premises usually in local and regional centres. Most European countries adopted this approach, but it was particularly strongly pursued in France, Ireland, Italy, the UK and across Scandinavia. In some regions the emphasis was on the development of tourism facilities, particularly around the Mediterranean, but also in remote and mountainous areas across central and northern Europe.

By the late 1970s the exogenous model of rural development was falling into disrepute (see Table 2). The continued intensification and industrialisation of agriculture came up against the saturation of domestic markets, against ecological limits (with rising problems of agricultural pollution and ecological degradation) and against a greatly diminished capacity in the urban sector to absorb the surplus rural population. Moreover, the recession of the early 1980s resulted in the closure of many branch plants and a growing sense that rural regions that had attracted a great deal of such inward investment were highly vulnerable to fluctuations in the world economy. Areas that had experienced rapid expansion of tourism also came to realise its seasonal and cyclical fluctuations as well as the destructive impact on local cultures and environments of mass tourism. Terms such as 'branch plant economy' and 'development without growth' were coined to highlight the incorporation of such regions within the global business logic of firms governed elsewhere; a logic working against any self-governing and self-sustaining regional economic development (Amin, 1993, p.2).

Table 2 Criticisms of Exogenous Approaches to Rural Development

- *dependent development*, reliant on continued subsidies and the policy decisions of distant agencies or boardrooms;
- *distorted development*, which boosted single sectors, selected settlements and certain types of business (e.g. progressive farmers) but left others behind and neglected the non-economic aspects of rural life;
- *destructive development*, that erased the cultural and environmental differences of rural areas;
- *dictated development* devised by external experts and planners.

Endogenous Approaches

These difficulties encouraged the exploration in the 1980s of so-called endogenous approaches to rural development (see Table 3) based on the assumption that the specific resources of an area - natural, human and cultural - hold the key to its sustainable development (Van der Ploeg and Van Dijk, 1995). Endogenous development ideas drew on four separate sources.

Table 3 Endogenous Approaches to Rural Development

Key principle - the specific resources of an area (natural, human and cultural) hold the key to its sustainable development
Dynamic force - local initiative and enterprise
Function of rural areas - diverse service economies
Major rural development problems - the limited capacity of areas and social groups to participate in economic and development activity
Focus of rural development
- capacity-building (skills, institutions and infrastructure)
- overcoming social exclusion

First there was the recognition that out of the economic restructuring of the 1970s and 1980s certain rural regions, with previously unrecognised internal dynamism, had emerged as leading economic regions. The Third Italy was the most celebrated example but successful rural regions could be identified across Western Europe, including, for example, East Anglia, Alsace, Bavaria, the Tyrol and South Jutland. The question arose of what was the key to success for these regions and whether it could be replicated elsewhere. Picchi (1994) cites the following elements as critical to development 'from within' in the Emilia-Romagna region of Italy: the importance of the agricultural sector for the provision of capital and labour needed in non-agricultural enterprises; the ability of this labour to engage in new economic activities; the cultural orientation towards self-employment; an extensive network of small- and medium-sized enterprises; and a dense system of interdependencies between economic sectors and units. He also identifies a set of political-institutional arrangements which have helped strengthen endogenous development patterns. These include a rich network of services provided by local administrations for economic sectors, economic planning mechanisms and a stable climate for industrial development.

The second source of endogenous development ideas was regionalist movements and agencies seeking to overcome previous policy failures and to promote forms of local development less dependent on external capital. The emphasis shifted to rural diversification, to bottom-up rather than top-down approaches, to support for indigenous businesses, to the encouragement of local initiative and enterprise and, where these were weak, to the provision of suitable training. Prominent examples of this kind of approach can be found in the activities of development agencies particularly in peripheral regions of Europe, for example in the Irish Gaeltacht, in the local contract plans drawn up in the fragile zones in France, in rural Wales, in mountain community projects in Italy, in village development groups in Northern Sweden and in local development agencies in the Austrian Alps.

The third source of endogenous development ideas was from the debate about rural sustainability. Increasingly, the environmental and natural resources of rural areas have come to be valued, and forms of development favoured that benefit from and enhance those resources. The sustainability concept seeks to bridge not only the conventional divide between economic development and environmental protection but also embraces the viability of localities and communities on which the maintenance of both the environment and economic activity ultimately depends (Redclift, 1991; Norgaard, 1994). Thus there has been a growing awareness that a conserved countryside must be socially viable and is therefore dependent on the vitality of rural communities (Lowe and Murdoch, 1993).

The fourth source of endogenous development ideas comes from notions of self-reliance promoted by two groups - radical greens and development activists working with particularly marginalised groups. The former have elaborated the 'small is beautiful' thinking of Shumacher into the field of community economics. The intention is to reassert local control over economic activities (Dobson, 1993). Douthwaite (1996) argues that sustainability requires communities to have control over their economies to protect themselves from the forces of globalisation and restructuring. This means that, where possible, local production should seek to supply local needs and strategies should be pursued for retaining value added from the use of local resources within the area. It also implies appropriate control over basic services, energy production and distribution (through alternative energy schemes), and finance (e.g. through credit unions, community banks, Local Exchange and Trading Systems, local currencies, etc.). Development activists working with marginalised groups have also promoted notions of self-reliance. A feature of community development in peripheral regions such as the West of Ireland, the Scottish Highlands and Islands, the mountainous areas of Italy, Southern Spain and Northern Sweden has been the promotion of community enterprises and community ownership and management of natural resources, through the formation of craft, fish farming, tourism, forestry and agricultural co-operatives (Hawker and Mackinnon, 1989; Varley, 1991).

3. The Evolution of European Rural Development Policy

The evolution of European policy has followed, incorporated and influenced these changing conceptions of rural development. In the 1960s and 1970s the dominant preoccupation was with agricultural modernisation. Since its inception in the 1960s, the CAP has included an agricultural structures component to assist the workings of the common market through a process of agricultural modernisation. This agricultural structures policy was funded through the Guidance section of the EAGGF — the European Agricultural Guidance and Guarantee Fund. It was primarily to help improve the production structures of agricultural holdings and the arrangements for processing and marketing agricultural products. In the 1970s, it took on also more of a social dimension with measures added to support early retirement and young farmer schemes, and for farming in Less Favoured Areas (LFAs). LFAs represented the first territorial component of agricultural structures policy. It recognised that the policy-led process of rationalising and consolidating farm structures might lead to the abandonment of some areas, and that the maintenance of farming communities in marginal areas was a desirable social goal.

The accession of Spain, Portugal and Greece to the European Community prompted a more fundamental rethink of the relationship between agricultural and rural development. Each of these countries had large peasant populations and clear needs for investment in their rural areas. But they added further to the Community's surpluses in the major Mediterranean products, at a time when Community decision-makers were becoming preoccupied with the problems of overproduction and mounting budgetary costs arising from the CAP.

Agricultural structures policy began to be shifted away from enhancing productivity to improvements in the quality of, and establishing new markets for, agricultural products, as well as beginning to develop environmental justifications and objectives for farming support. At the same time it was apparent that the modernisation of the agriculture of Southern Europe would need to be accompanied by large-scale investment in the infrastructure, services, and non-agricultural sources of employment of rural regions. This was part of the thinking behind the major reforms and expansion to the Structural Funds in 1989 and 1993. The new Objective 1 designation was intended to channel European funding into the regional development of the so-called cohesion countries – Spain, Portugal, Greece and Ireland. At the same time it was recognised that certain agriculturally dependent regions within the rest of the Community would need assistance to begin to diversify their economic base away from uncompetitive farming activities. This was the intention of the new Objective 5b designation. The wider reforms of the Structural Funds sought to combine the different funds (the EAGGF, the European Social Fund and the European Regional Development Fund) in regionally targeted and co-ordinated programmes. As part of this process there was a major expansion of the Guidance section of EAGGF most of which became part of a broader territorial approach to integrated development, with new partnership and decision-making arrangements for programme management put in place between the European Commission, Member States, and sub-national actors.

In a seminal Green Paper, published in July 1988, the European Commission had set out a strategic re-think of its rural policy titled The Future of Rural Society. The message of the need for an approach that stimulates development from within came through strongly in the report. The Commission explained that rural development policy "must ... be geared to local requirements and initiatives, particularly at the level of small and medium-sized enterprises, and must place particular emphasis on making the most of local potential". This does not, the Commission argued, mean "merely working along existing lines. It means making the most of all the advantages that the particular rural area has: space and landscape beauty, high-quality agricultural and forestry products specific to that area, gastronomic specialities, cultural and craft traditions, architectural and artistic heritage, innovatory ideas, availability of labour, industries and services already existing, all to be exploited with regional capital and human resources, with what is lacking in the way of capital and co-ordination, consultancy and planning services brought in from outside". This thinking informed the development of Objective 5b but more particularly of LEADER.

Since the early 1990s, two alternative visions have competed for influence in determining the evolution of EU agricultural and rural policy. On the one hand, 'market liberalisers' have pressed for reductions in commodity prices, and the removal of export aids to open up the European agricultural market to world trade. On the other hand, the protectionists have argued that such moves would be seriously detrimental to farming communities across the European Union and have resisted calls for further CAP reforms. Between these two positions, a 'third way' has emerged, pressing for progressive reform of commodity supports to be linked to the promotion of an integrated/rural development agenda.

In 1995, agricultural Commissioner Fischler opened up the prospect of further CAP reform in the light of the need to accommodate EU enlargement to the east, and the next round of World Trade Organisation talks on agriculture. In November 1996, he convened a conference on rural development at Cork in Ireland to try to widen popular support for his ideas on CAP reform and the building of an 'integrated rural policy'. Central to the Cork agenda was the notion that some element of the resources saved from future reductions in commodity price support and direct payments under the CAP should be recycled within rural areas through agri-environment, agricultural structures and rural development spending. The 'Cork Declaration' pointed towards an expanded rural development programme — along similar lines to Objective 5b and LEADER — with an emphasis on including the whole, farmed countryside within its scope, rather than focusing on specific geographical zones (see Table 4 overleaf).

The Commission's formal *Agenda 2000* proposals were subsequently issued in July 1997. The *Agenda 2000* document contained outline proposals to move the CAP away from an emphasis on production support and towards a more 'Integrated Rural Policy' combining subsidies to farmers — decoupled from production — with rural environmental management and economic diversification measures. A more detailed package of proposals was published in March 1998 in the form of draft regulations. The Explanatory Memorandum accompanying the regulations referred to rural development (including agri-environment schemes) becoming *"the second pillar of the CAP"*. However, the proposed resourcing of this second pillar disappointed environmental and rural development interests. It was allocated some 14% of total CAP resources, although when spending on Eastern European applicant states was excluded from the calculations, the proposed spending in the EU15 would effectively fall slightly from 1999 levels to approximately 10% (see Table 5 overleaf).

The Berlin Summit in March 1999 saw the conclusion of the negotiations on *Agenda 2000*. The resources available for the new Rural Development Regulation (RDR) were widely regarded as disappointing. A significant degree of national discretion, though, was introduced in the deployment of some of the direct payments to farmers, in relation to national and subnational objectives for the countryside, allowing an unprecedented tailoring of the CAP and the transfer of additional resources to the RDR. The arrangements for programming and implementing the RDR, moreover, represent a new model imported from the operation of Structural Fund programmes. Each Member State is required to draw up territorially-based seven year rural development plans "at the most appropriate geographical level" initially for the period 2000-2006. All rural development and agri-environment support measures are to be integrated within a single plan. Although the former measures are optional, and a very broad-based menu of measures is allowable, the latter are compulsory with the result that each region will have to have an agri-environment programme in accordance with its specific needs. The Structural Funds will continue to be complemented by a Community Initiative on rural development, with LEADER+ continuing to develop the approach of past LEADER programmes. SAPARD is the equivalent of the RDR for the CEECs.

Table 4   Cork Declaration

The Main Principles - A Summary
1. **Rural Preference.** Sustainable rural development must be at the top of the EU agenda and become the fundamental principle that underpins all rural policy.

2. **Integrated Approach.** Rural policy must encompass within the same legal and policy framework agriculture, economic diversification, management of natural resources, environmental enhancement and the promotion of culture, tourism and recreation. It must apply to all rural areas in the Union and be subject to co-financing differentiated according to need.

3. **Diversification.** Support for diversification of economic and social activity must focus on providing the framework for self-sustaining private and community-based initiatives.

4. **Sustainability.** Policies should promote rural development which sustains the quality of Europe's rural landscapes, including natural resources, biodiversity and cultural identity.

5. **Subsidiarity.** Rural policy must be as decentralised as possible within a coherent European framework that ensures partnership between all the levels concerned (local, regional, national and European). The emphasis must be on participation and a 'bottom up' approach to harness the creativity and support of rural communities.

6. **Simplification.** Rural policy, especially its agricultural component, needs radical simplification of legislation to ensure greater coherence, subsidiarity and flexibility, but without a renationalisation of the CAP.

7. **Programming.** The implementation of rural policy should be through a single integrated programme for each region.

8. **Finance.** Local financial resources must be encouraged to promote rural development, with improved synergy between public and private funding.

9. **Management.** The capacity and effectiveness of regional and local government and community groups must be enhanced, through improved training, technical assistance, networking and exchange of experience.

10. **Evaluation and Research.** Monitoring, evaluation and beneficiary assessment will need to be reinforced

Table 5: The Changing Architecture of the CAP

1990	MacSharry Reforms 1996	Agenda 2000 Reforms 2002
Market Support	Market Support	Market Support
	Compensation Payments	Compensation Payments
	Agri-environment	National Envelopes
	Structural	Rural Development Regulation

Funds recouped from imposition of Cross-compliance and Modulation

27

4. Rural Economic Change

These policy changes are having to adjust to and tackle rapid economic changes that are transforming the functions of rural areas and setting them on to quite different development trajectories. On the one hand, there has been an inexorable decline in primary sector employment and traditional rural industries have been squeezed. On the other hand, new industrial and service activities have emerged, although not necessarily in those regions suffering the most from rural decline.

All the EU countries have suffered losses of primary sector employment over several decades. There are now few regions in the EU where agriculture contributes more than 10% of the regional value added and these are concentrated in Greece, Portugal and Ireland. Agriculture now accounts for only 5 per cent of employment in the EU (Eurostat, 1998). Forces of mechanisation have widely affected not only agriculture, but forestry, fishing and mining too; and expansion of production has encountered problems of over-exploitation and over-supply. At the same time, processing and manufacturing activities once closely linked to the primary sector (such as farm machinery manufacture, food processing, the leather industry, timber processing, etc.) have undergone significant economic and geographical concentration and face growing competition from outside the EU. Many service activities traditionally found in rural centres have also experienced intensified competition from urban centres. The consequence of all these developments has been the loss of much localised employment from rural areas and regions.

At the same time, new economic functions have emerged for rural areas. Indeed, new firm formation rates and employment growth have tended to be higher in small towns and rural areas than in large urban centres. There has been a net increase in employment in all non-metropolitan regions of the EU with the exception of Greece and Finland (OECD 1996). In some cases growth is due to the decentralisation of productive activities, but very often it is due to indigenous industrialisation. Furthermore, in more central regions, certain service activities have also relocated to rural areas, thereby accentuating an employment pattern already heavily weighted towards the service sector, including leisure and tourism as well as public and social services.

In the CEECs in contrast, not only has agricultural production declined absolutely (not just relatively) but this has occurred without the compensating growth of rural industries and services. Indeed, in many cases what was there before has since collapsed. In the CEECs, agricultural systems have experienced dramatic changes since the late 1980s. The dismantling of state structures of management and control, the privatisation and restitution of landownership and the sweeping market reforms of the early 1990s massively disrupted agricultural production in the short term and have greatly diminished the intensity of production in the medium term. The decline in output was most pronounced in livestock production as consumers switched to cheaper staple products and export markets were lost. In most countries, cattle and sheep numbers fell to about half their former level and there was a decline of 30-35% in pig and poultry populations. Crop production fell by up to a third compared to 1989 but there has been an increase in average yields and production in most countries recently. Overall a great deal of capital and labour have been withdrawn from the agricultural sector. Rural unemployment levels are high. There is also extensive land abandonment, particularly in regions where growing conditions are poor and which are peripheral (at the same time the geography of peripherality is changing as markets switch from East to West). These developments have been exacerbated in some cases by the way the privatisation of land has been handled. Privatisation of the state owned and collective farms generally has resulted in a dual farm structure. Typically, there is a large number of small semi-subsistent family farms, subject to rapid amalgamation in some areas, which exist alongside a group of larger units comprising co-operatives, limited companies and state farms.

5. Comparative Dimensions to Rural Development Policy in Europe

National perspectives on rural development are thus driven by quite different sets of concerns. The range of prevalent concerns includes the following: agricultural modernisation; infrastructure development; maintaining regional populations; landscape protection and management; rural economic diversification;'social and economic cohesion; and relieving rural disadvantage and deprivation. Comparing national agendas, a number of key axes can be discerned:

An Agrarian versus a Rural Perspective

One very important axis is an agrarian versus a rural perspective. Many would not recognise the distinction, but others would see it as fundamental. Some would see rural development as an adjunct to agricultural policy; others would see agricultural development as a component of rural policy. Beyond simply confusion over terms, at the core of these disagreements is a debate about the continuing centrality of farming, socially, culturally and environmentally, to the future of rural areas and national societies.

In all EU countries, agriculture represents an ever-diminishing proportion of GDP and employment. In most countries, though, it is still the major land use shaping the national territory and regional geography, forming the rural environment and acting as a reservoir of cultural values. Agricultural policy was clearly once seen as a means to support rural areas. Conversely, some now promote rural policy as a means to support agriculture and farming communities, while others see rural policy as a means to help rural areas overcome their dependence on a sector (agriculture) that is in decline and move towards a post-agricultural countryside. Relating to these alternative viewpoints are questions concerning the legitimacy of continued public funding for farmers as guardians of the countryside, the significance accorded other economic and political actors in rural development strategies, the relative importance given to farming and non-farming activities, attitudes towards farm structural change, views about desirable alternative activities for rural areas, whether to diversify farming or diversify the rural economy, etc.

There are different perspectives on these issues even within countries, but broad national differences are also apparent. Farming lobbies and agrarian ideologies are more powerful in certain countries than in others.

For example, the central preoccupation of French rural policy has always been the prospects for agriculture and its changing role in rural areas. Likewise Austrian rural policy is built on the supposition that agriculture and rural areas determine each other and there is an increasing homogenisation of agrarian and local development perspectives on rural policy. In Germany, the GAK which is the main instrument to support rural areas is legally limited by the Basic Law to the improvement of the agricultural structure; one implication is that business support schemes for the countryside are restricted to processing and marketing of forestry and agriculture and are not available for small business and crafts.

In contrast, in the UK there has been an historical separation of a rural development policy, preoccupied with rural industries and village services, and a sectoral agricultural policy. Likewise, in Sweden agriculture constitutes only a minor part of rural development : at least as important are forestry, fisheries, reindeer husbandry, mining, hydropower and small enterprises of various kinds. Finally, in a number of the CEECs, including Hungary, much of the rural population, although now subsistent farmers, used to be industrial workers, and it may be quite inappropriate to cast their futures in an agrarian perspective

Traditional Economic Development versus Consumption/Recreation
Traditionally rural areas have been seen as areas of productive activity and places of work. Increasingly, though, they have come to be appreciated as places to live and sites for leisure. This transition clearly depends upon levels of affluence and the spread of post-materialist values in society. As people move beyond concerns with material security and embrace quality of life issues they place increasing value on the opportunities rural areas provide for living space, self-expression, recreation, the enjoyment of amenity and wildlife, and a wholesome and pleasant environment.

These tendencies are most marked in the central, most advanced economic regions of the EU which have large, middle-class commuter belts, but also in attractive peripheral areas which have developed functions for tourism, second homes, retirement and nature protection. In addition, there are different cultural norms operating - the consumption countryside in England is about landscape appreciation and recreational access; in France or Italy it is more about gastronomy and wine; in Austria and Sweden it is about lifestyle and community.

While a common policy framework for rural Europe must increasingly accommodate this post-modern consumption countryside, regions of Southern Europe still present classic problems of rural underdevelopment, while the CEECs have large rural populations suffering extensively from impoverishment and immiserisation. The preoccupations of rural development institutions thus range from the traditional, such as the need to increase employment, reduce rural poverty and improve infrastructure, to a newer agenda which includes building social capital, tackling gender imbalance, promoting cultural expression, heritage protection, supporting alternative forms of agriculture and improving monitoring and evaluation. In the more affluent northern countries, such as Austria, Germany, Denmark, Sweden and the UK the second set of concerns has gained more ground. However, elements of this debate are also present in Southern Europe and even in central Europe, alongside the more pressing concerns of large scale unemployment, land abandonment and underinvestment in infrastructure, education and training.

6. Conclusions: the challenges posed by the Europeanisation of rural policy

At the European level, what constitutes 'rural development' has come to be understood in recent years as those activities eligible for support under the CAP's new Rural Development Plans across the whole countryside, LEADER and the new Structural Fund programmes in rural areas designated as within Objective 1 and Objective 2 regions. A parallel process is underway in the CEECs through their SAPARD programmes. This is, of course, a highly particularised version of 'rural development' which reflects the currently prevailing concerns of those responsible for EU agricultural, cohesion and accession policy. It differs to some degree from other notions of rural development which have either characterised EU policies of the past, or which continue to characterise national or local understandings of rule development in different parts of Europe.

Comparing the current 'European model' of rural development with those of the past, we can see that what now constitutes 'rural development' has widened significantly to embrace the agri-environmental measures first introduced under the 1992 CAP reforms. Prior to Agenda 2000, agri-environment policy and rural development policy were far less integrated, with the latter consisting of agricultural structural measures and the rural parts of Objective 1 and 5b programmes. Now these two policy areas have come together under the Rural Development Regulation. On the one hand this process could be interpreted as a broadening of the scope, and the Europeanisation of the scale, of rural development policy. However, on the other hand, the fact that Member States are able to draw up their own programmes from a menu of measures means that what constitutes rural development still has the scope to vary, within this framework, from place to place. It is vital that the new European model of rural development be flexible and inclusive, given the continued diversity of rural problems and institutional definitions of rural development across Europe.

The Rural Development Regulation and the SAPARD, however, are rooted in the evolution of agricultural structures policy. They are still predominantly preoccupied with agricultural adjustment and have a fundamentally agrarian perspective on rural development. Quite apart from their inadequate resourcing, they are limited vehicles with which to address the non-agrarian aspects of rural development, either as manifested in the post-agricultural and consumption countryside of Western Europe, or in the large-scale rural restructuring required across Central Europe and in parts of Southern Europe.

References

Amin, A. (1993) *The regional development of inward investment in the less favoured areas of the European Community* Paper presented at the Conference on Cohesion and Conflict in the Single Market, Newcastle upon Tyne.

Clout, H. (1993) *European Experience of Rural Development* London: Rural Development Commission.

Dobson, R.V.G. (1993) *Bringing the Economy Home from the Market* Quebec: Black Rose Books.

Douthwaite, R. (1996) *Short Circuit - Strengthening Local Economies for Security in an Unstable World* Dartington: Green Books.

Eurostat (1998) *Labour Force Survey - Results 1997* Luxembourg: Eurostat.

Hawker, C. and Mackinnon, N. (with Bryden, J., Johnstone, M. and Parkes, A.) (1989) *Factors in the Design of Community Based Rural Development in Europe* Aberdeen: Arkleton Trust and Scottish Development Association.

Lowe, P. and Murdoch, J. (1993) *Rural Sustainable Development: report for the Rural Development Commission* London: RDC.

Norgaard, R.B. (1994) *Development Betrayed: the End of Progress and a Coevolutionary Revisioning of the Future* London: Routledge.

OECD (1996) *Territorial Indicators of Employment - Focusing on Rural Development* Paris: OECD.

Picchi, A. (1994) 'The relations between central and local powers as context for endogenous development' in Van der Ploeg, J.D. and Long, A. (eds) *Born from Within: the Practice and Perspectives of Endogenous Rural Development* Assen, The Netherlands: Van Gorcum.

Redclift, M. (1991) 'The multiple dimensions of sustainable development' *Geography*, 76 (1) pp.36-42.

Van der Ploeg, J.D. and Van Dijk, G. (eds) (1995) *Beyond Modernization: The Impact of Endogenous Rural Development* Assen, The Netherlands: Van Gorcum.

Varley, T. (1991) 'On the fringes: community groups in rural Ireland' in Varley, T., Boylan, T.A. and Cuddy, M.P. (eds) *Rural Crisis: Perspectives on Irish Rural Development*. University College Galway: Centre for Development Studies.

Rural development and the environment:
How to remunerate farmers for their contribution ?

Guido Van Huylenbroeck

Department of Agricultural Economics, Ghent University, Coupure Links 653, 9000 Gent, Belgium

Introduction

There is no doubt that the actual state of the European countryside is highly influenced by the use of rural areas for the production of Food and Fibre (F&F). About half of the land in Europe is under agricultural production (European Commission, 1999). As long as agricultural operations were relatively small in scale and using proper technologies without much external inputs, this did not cause major problems. On the contrary, the economic value for the rural population and landowners of maintaining and improving the resource base of the land by carrying out stewardship practices such as maintenance of roads and hedges, drainage and water regulation systems was a sufficiently high incentive to create and maintain specific landscapes, which are now regarded as the norm and highly valued. Also the fact that most people in the rural areas were living from agriculture created a harmony between production and conservation issues.

The introduction and increased use of external inputs (agro-chemicals and compound feed) and mechanisation of agricultural practices has reduced in the past 50 years the economic value of countryside stewardship. Landscape elements became less functional and even a hindrance to intensive agricultural practices. This has caused two related but highly different problems: on the one hand intensification of agriculture in those areas best suited for agricultural production resulting not only in the removal of landscape elements, but even in the destruction of the environmental resource base (pollution) and on the other hand the problem of land abandonment in areas where a competitive form of agriculture is no longer remunerative resulting in a degradation of these ancient man made rural areas.

In both situations economic reasons are the cause of the environmental degradation. Therefore rural development, countryside management and environmental protection have to go hand in hand. Their objectives are not contradictory because stimulating the production of environmental goods and services can create new economic opportunities and thus contribute to a diversification of rural activities, while on the other hand innovative rural activities such as production of quality or regional products, rural tourism and so on can contribute to countryside management as they depend highly on the quality of the rural environment (Leon, 1999). This has been well understood in Agenda 2000 where rural development and agri-environmental policies are brought under the same umbrella. The main question is then how to stimulate farmers in these activities because most farmers will not contribute to quality production, environmental protection of landscape maintenance as long as markets do not remunerate these practices that result in higher costs. It is then the role of policy to design and introduce systems that make farmers aware of their countryside management function and deliver incentives which will persuade them to change their practices.

In the remainder of this paper first the reasons why public intervention is necessary are reviewed, after which the main mechanisms that can be used to remunerate farmers for their contributions are given. The paper continues with the main results of a study on actual applied countryside stewardship policies in eight EU-countries (see Van Huylenbroeck and Whitby, 1999) and further discusses some governance structures for intermediate organisations.

Reasons for public intervention

The main reasons for public intervention have to be found in the economic nature of environmental goods (EGS) that can be characterised by the following words: character of this impact.

The role of public intervention is therefore the reconciliation of the balance between the F&F output and the production of Environmental Goods and Services (EGS). In Fig. 1 this relationship is illustrated. When agricultural intensity increases, EGS-production decreases.

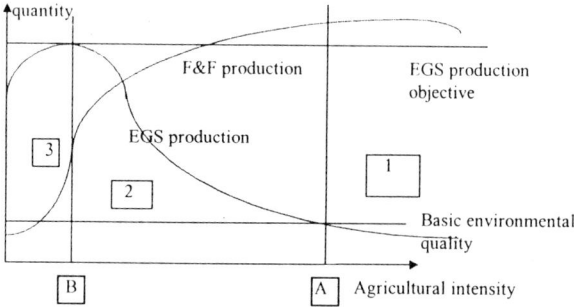

Fig. 1: Schematic relationship between F&F and EGS production with increasing intensity of agriculture

Three zones can be distinguished. In zone 1 (beyond point A) agricultural intensity is causing pollution or unacceptable effects on the landscape. In this case binding and mandatory policies limiting the individual property rights - which are at that moment in conflict with the common property rights on pure air, water or pleasant countryside – have to be applied to avoid the unacceptable impact.

In zone 2 (between point A and B), agriculture is respecting basic environmental quality. If in this case a higher EGS provision is desired, economic incentive instruments changing the economic conditions in favour of lower intensity need to be applied. Finally, in zone 3 EGS-production becomes the primary objective (e.g. in nature reserves or highly valued zones). Here, agricultural activities can be accepted as far as they support the development of natural values. In this case shifts of property rights to public authorities and policies preventing land abandonment become more appropriate.

Public intervention mechanisms

The first and maybe most important task of public intervention is to fix the exact location of the reference points A and B of Fig. 1 and to make them respected by use of juridical and legal binding instruments which must on the one hand separate situations considered as pollution from non-pollution by issuing standards that need to be respected and on the other hand design vulnerable zones in which property rights are limited by planning instruments and licences (for larger zones) and land purchase by public bodies (for smaller zones). As indicated by Gatto and Merlo (1999) the definition of these reference points will depend on local conditions, on the features of the particular EGS to be protected (with internal conflicts between different externalities) and on social and political tradition and property regimes in a country. In between point A and B there is scope for economic instruments.

However to be effective also regulatory policies need often to be accompanied by financial or economic instruments. In order to make farmers comply with standards, a system of taxes or levies should be used. Compensatory payments are justified in cases where designation causes substantial modifications of property rights while incentives should be used to persuade farmers to enter in a scheme with positive contributions for the countryside or to farm marginal land. In theory all financial or economic instruments are of a voluntary nature at least if possibilities exist to avoid the taxation, possibly by switching activities, or to refuse the compensation by not entering in a scheme. In practice a lot of taxes, levies or even economic compensations have a mandatory character as the only possible escape is not to produce or to cease activities. Such regulatory policy is however only possible when public authorities have full information about the causes of degradation and can fully control the actions of individual actors. If this is not the case a less regulative but more stimulating policy using market-led instruments is more appropriate (Mormont and Van Huylenbroeck, 2000).

Indeed, a third possibility of providing environmental goods is through market-led instruments. These are real voluntary policies for which two types can be distinguished: creating a market by the state offering payments or by creating conditions for the establishment of a market for private goods. To the first category belong instruments like management agreements or the auction of incentives. Both instruments are based on the creation of a market through which compensatory or incentive payments for the delivery of EGS may be negotiated. Management agreements are based on individual negotiations between the public authorities and the farmer in which the authorities offer certain payments for certain countryside stewardship practices. The farmer may choose from the menu, the scheme most convenient to him. In an auction market, all interested farmers can bid to obtain the permit or licence to use an amount of land under predefined constraints or conditions (e.g. in the case of nature reserves). This requires that public authorities or a private trust is able to obtain the property rights.

A real market is established when the EGS can be sold to the consumer (beneficiary-pays-principle). The role of public intervention is then mainly restricted to the creation of conditions in which these markets can be established. Because by definition public goods have no price, this is only possible if the public good can be transformed in or linked with a private good. This is possible if private property rights can be assigned (e.g. access rights) so that someone (the state or private users) have to pay to be able to enjoy the public good (direct market) or if a joint product can be marketed (indirect market). In this case the marketable output receives a higher value when it is produced in combination with the EGS (a well known example is the Parmesan cheese). Labelling (e.g. AOC) or the creation of regional or typical products with specific quality characteristics are one possibility, establishment of a recreational market linked to the production of EGS another. The use of the land market is theoretically another market possibility. If the EGS becomes scarce, this can be reflected in the prices of the land (e.g. land prices in the neighbourhood of beautiful scenery), making the market mechanism start to work, but because of irreversibility this in most cases is too late because the EGS have disappeared and can not be produced again. Therefore it is rather difficult to use the land market mechanism actively to encourage EGS production.

Finally, persuasive instruments may be used. It is clear that through education, extension and publicity farmers can be made aware of their role. Also social pressure is important in making farmers change their behaviour. The fact that in recent decades public opinion has given greater importance to the environmental issue, makes (some) farmers take it more seriously. Another more drastic way of persuasion is cross-compliance. This means that the payment of other subventions is made conditional on the compliance with certain environmental requirements. In the Agenda2000 this possibility is to a limited extent introduced. Another alternative approach, belonging to the persuasion instruments is the use of covenants (Whitby, 1998). These are long term agreements between the State and a sector, or group of land owners or tenants to achieve certain environmental objectives without specifying the instruments. If the goal is not reached either a penalty clause comes into operation, or the State has the right to take more drastic measures. This possibility is already used e.g. in the Netherlands where agri-environmental payments are distributed through co-operatives of farmers.

Inventory and classification of existing countryside stewardship policies

In a recent EU-project (the so called STEWPOL-project) the countryside stewardship policies applied in the 8 participating countries have been inventoried and classified. Among the analysed policies where both agri-environmental schemes as programmes more belonging to the rural development policy (as e.g. certain LEADER programmes). The inventoried policies have been classified based on a set of policy indicators (see Bonnieux and Dupaz, 1999). Among the indicators were two groups of criteria measuring the performance of programmes or policies with on the one hand indicators describing if environmental objectives are reached and on the other hand economic efficiency criteria measuring if these goals are achieved at minimum costs for society. The third category of criteria tries to indicate how far policies are successful (in terms of uptake) and sustainable (in terms of having intended effects in the long term), while the objective of the fourth category of indicators (enforceability) was to include into the analysis the reliability of achieving objectives and the costs of administrative implementation and control. The fifth and sixth category of indicators focus on the political acceptability of measures and the compatibility with general policy principles such as the polluter-pays-principle. Most of the indicators, except for readily available figures like remuneration level or participation rate were mainly simplified indicators, showing whether a certain criterion is fulfilled or not, because real quantifiable measurements in particular of the environmental performance of policies are still rare or if existing not applicable because of lack of data.

Most of the policies inventoried belong to the category of voluntary schemes (94.4 %) situated in zone 2 of Fig. 1. This means that they try to change economic substitution conditions between EGS and F&F-production, mainly by offering compensatory payments or incentives to farmers for positive contributions to the environment or landscape or for the reduction of negative externalities not regarded as pollution, although almost 40 % of the applied policies violate to a certain extent the polluter-pays-principle by providing compensations for the reduction of unacceptable amounts of pollution.

A second observation from the inventory is that most CSPs have more than one objective with reduction of negative impact of agricultural practices as the most prominent one, but also conservation of landscape, environment and wildlife are important objectives showing that most of the policies have in mind (at least partly) a contribution to positive externalities. A lot of the inventoried policies are targeted to specific regional problems (vertical measures applicable only in specified zones), but are mainly financed and administrated at a higher level.

Most CSPs put restriction on agricultural practices (273 of the 355 identified policies) and/or require specific landscape conservation practices (maintenance of hedges, trees, etc.) (218 cases out of 355. The average remuneration obtained per ha is low (in average only about 100 euro per ha). It concerns mostly flat payments (54 % of the cases) or modulated according to the applied zones (20 % of the cases). In 37.8 % of the cases individual agreements are made but the level of compensation is only in a limited number of inventoried policies negotiated with the beneficiaries (7 % of policies). Real market-led instruments are only used in a limited number of CSPs although the produced EGS output is in 18 % of the cases already marketed and in 30 % of the cases potentially marketable. This means that in these cases use-values are produced while for the other half of the policies the objective is the provision of real public goods providing non-use or option values.

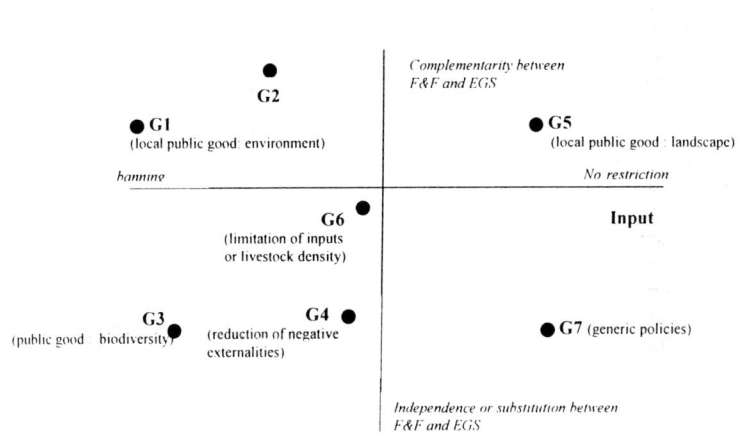

Fig. 2: Characteristics of the kernels of CSP-groups identified by correspondence analysis

Based on these indicators a multiple correspondence analysis has been used (Bonnieux and Dupraz, 1999) to reduce the data set, to identify discriminating factors and to arrive at a typology of the analysed CSPs. This analysis allows to identify among the inventoried policies seven groups or kernels which are characterised in Fig. 2. according to two factors namely the kind of EGS they intend to produce and the level of required limitation for external inputs. Above the horizontal axis, those policies are situated belonging to what can be characterised as rural development policies. They are policies trying to stimulate diversification of agriculture and/or stimulating so called multifunctional agriculture. These policies stimulate farmers in green tourism based on the provision of local environmental amenities attractive to people wanting to walk or cycling in the countryside (G1), in quality production (G2) by using labels of regional origin, organic production and so on or in the maintenance of landscape goods (like hedgerows, stonewalls, ...) (G5). The required practices are compatible with agricultural production as the outputs of these policies are potentially marketable and can create new activities in the rural area. However if they are actually marketed often depend on a certain organisation of farmers and the ability to create a separated market for the products.

The programmes belonging to the categories under the horizontal axis are mainly environmental policies reducing negative externalities of agriculture or stimulating nature production. These policies reduce to a certain extent agricultural activity as they are based on a reduction of these activities in return for a compensatory payment. As they mainly create intrinsic environmental value, they are hardly directly marketable. At the left side of the vertical axis are situated those policies requiring limitations on the inputs, while at the right side these limitations are hardly present as they are more based on payments for certain services such as maintenance of landscape elements or specific natural sites.

Costs of policies

The compensations to farmers are not the only costs of agri-environmental or rural development policies. The running of these policies requires important functioning expenses in order to install, monitor and control the envisaged actions. These costs are in economic terms characterised by the term transaction costs. These include all those costs that are necessary to make sure that socially-desirable transactions take place. Public provision is only justified when total costs are lower than under market mechanisms or when market transactions do not take place. But also then analysis of transaction costs of public policies can be interesting to detect inefficiencies in management and design of policies, in particular if their existence or amount can not be explained or justified.

In their analyses Falconer and Whitby (1999) show that transaction costs (including information, contracting and policing costs) for CSPs are in general higher than for market control and price policies (> 25 % versus less than 5 %). A more detailed analysis of a number of inventoried policies (42) has allowed further analysis of the existence and differences in transaction costs. Without going into detail, the analysis reveals that transaction costs tend to decrease in time (due to the recovery of high starting costs of new policies), tend to be lower for schemes with high participation rates (economics of scale), tend to be higher the more vertical and specifically targeted policies are (because of higher control costs, but a trade-off exists with saving on compensations to be paid when policies are better targeted). The fact that control for a number of the inventoried policies is rather difficult and not very reliable is an important issue in this respect. Further analysis reveal striking differences indicating the importance of administrative structure for transaction costs as well as the existence of scope to decrease them. Suggestions to decrease transaction costs can be found in better organisation of administration and control, co-ordination of administrative files per farm (and not per programme as is the case now) or privatisation (of part) of the required administration.

However, in the long run self-organisation and the use of market mechanisms will be less expensive. However, also then transaction costs can be of importance to choose the most appropriate governance mode. As will be shown in the next section

Governance structures for intermediate organisations

When direct or indirect delivery through public intervention is too expensive or not possible because the mechanisms are not known or difficult to control, the delivery through intermediate organisations can be a good solution. These intermediate organisations can help to construct a market for specific products and thus support the use of market mechanism in the delivery of environmental goods and services. Examples of such intermediates are farm markets, food teams, food delivery organisations, co-operatives for processing and marketing farm products, environmental co-operatives, organisations selling regional products and so on. Through these intermediate organisations so called club goods can be realised resulting in common property rights making market separation possible. Those who are members of the organisation can take advantage of the benefits of the joint effort (in return for a membership fee or doing part of the work involved) while others can be excluded (creating excludability and rivalry).

Depending on the specific assets involved, the frequency of transactions, the uncertainty in the market, different economies on transaction costs can be realised through different governance structures. From a research for a number of innovative marketing channels for specific food quality, that the more specific the assets necessary for producing or processing the specific quality, the more integrated the collaboration has to be (Verhaegen and Van Huylenbroeck, 2000). Benefits for farmers come from the fact that they have not to invest themselves in these specific assets (such as e.g. time to be devoted to marketing activities, processing material, publicity, specific knowledge and so on) and they can reduce marketing risks and increase the volume that can be sold. On the other hand the organisation needs to be stronger because the more specific assets, the more dependent the organisation is on the supply and respect of rules and practices by farmers making necessary more formal contracts and safeguards to protect the interests of all partners involved. This explains why farmer markets or food teams can have a rather loose governance structure as in these organisations only few specific assets are involved (every farmer remains responsible for his own production, processing and marketing) while organisations building up common knowledge, selling under regional label, organising common processing or rural tourism or doing other activities in common (as e.g. maintenance of landscapes) need a stronger central organisation.

For the State the advantage of such intermediate networks is that its transaction costs will decrease (as they have not to negotiate with and to control every individual farmer), part of the costs for rural development or environmental protection will be paid by the consumer (at least if marketable products are realised). A global welfare increase can be realised if the total delivery cost is lower than under other governance structures.

Conclusions

In this paper an attempt has been made to review the role of public intervention in rural development and the provision of environmental goods and services. It has been argued that rural development and the delivery of environmental goods and services go hand in hand and can reinforce each other. Public intervention is necessary because it concerns often public or common pool goods which are not delivered through market governance. In order to provide these goods the State can use different mechanisms going from own delivery, command-and-control procedures, economic instruments to stimulating delivery by self governance. Depending on the total cost of each governance mode as well as on the available knowledge and control possibilities of public authorities, the best policy has to be chosen. In practice we can observe a shift from the "regulating" State to the more "stimulating" State setting conditions in which private and collective action become possible. In particular, stimulating collective action opens a lot of possibilities with respect to rural development and environment. However, this requires institutional change at all levels.

Another observation with respect to rural development and the delivery of environmental goods and services, is the importance of targeting policies to the needs of the region and the prevailing environmental problems. Diversity in Europe is so large (see European Commission, 1999) and environmental problems so different, that no sole solution exists. Even at regional scale differences in the role of agriculture and needs for development can be high as demonstrated for Flanders in a recent study (Vernimmen and Van Huylenbroeck, 2000), that policies need to focus on local problems and opportunities. Therefore policies need to be sufficiently decentralised so that within general rules local accents are possible. Also this requires changes at the institutional level.

Finally, given the subject of this conference on higher education, it is important to emphasise that also at educational level these developments require new approaches: students in agricultural economics need much more than is the case to be confronted with new economic theories which are very useful to understand reasons behind certain developments in rural areas such as the new institutional economics approaches with *inter alias* the transaction cost and principal agents theory, convention and innovation economics, social network theory, etc. Formation in rural development also requires knowledge about other economic activities in the countryside such as tourism, retail selling of farm products, small and medium businesses, and so on. Maybe the time has come to change the name and content of the discipline from agricultural economics to rural economics !

References

Bonnieux, F. and Dupraz, P. (1999) Policy indicators and a typology of instruments. in Van Huylenbroeck, G. and Whitby, M. Countryside stewardship: farmers, policies and markets. Oxford, Pergamon press, pp. 47 – 66.

European Commission (1999) Agriculture, environment, rural development: Facts and figures – A challange for agriculture. Brussels, European Communities, 261 p.

Falconer, K. and Whitby, M. (1999) The invisible costs of scheme implementation and administration. In Van Huylenbroeck, G. and Whitby, M. (eds) Countryside stewardship: farmers, policies and markets. Oxford, Pergamon press, pp. 67 – 88.

Gatto, P. and Merlo, M. (1999) The economic nature of stewardship: complementarity and trade-offs with food and fibre production. Oxford, Van Huylenbroeck, G. and Whitby, M. (eds) Countryside stewardship: farmers, policies and markets. Pergamon press, pp. 21 – 46.

Leon, Y. (1999). The economic Analysis of Rural Development. Paper presented at the IXth European Congress of Agricultural Economists, Warszaw, August 24-28, 1999.

Van Huylenbroeck, G. and Whitby, M. Countryside stewardship: farmers, policies and markets. Oxford, Pergamon press, 232 p.

Mormont, M. and Van Huylenbroeck, G. (2000) A la recherche de la qualité: analyse socio-économique des nouvelles filiéres agro-alimentaires (forthcoming).

Verhaegen, I & Van Huylenbroeck, G. (2000) Analysing farm diversification activities with the transaction cost theory. In Hillebrand, H.; Goetgeluk, R. & Hetsen, H. (ed). Plurality and rurality: The role of the countryside in urbanised regions. The Hague, Agricultural Economic Institute, p. 27-40.

Vernimmen, T. and Van Huylenbroeck, G. (2000) The hidden countryside in Flanders. Paper presented at the EAAE-seminar on rural development in Aberdeen.

Whitby,M. (1998) Agri-environmental policies in the UK: a review of experience. Paper presented at a seminar in Lisbon organised by the Portuguese Ministry of agriculture.

Rural areas in transition: the sociological perspective.

Andrzej Pilichowski

Institute of Sociology, University of Lodz, PL 90-214 Lodz, 41 Rewolucji 1905 r str,.Poland.

Introduction

The topic of this year's Conference affords a good opportunity for starting this presentation with making reference to two fundamental issues, that is, time and space. The Conference Subtitle: Challenges for Higher Education in the New Millennium is an inspiration for the author to resort to a development perspective. Changeability of social forms is an obvious phenomenon for the sociologist. At crucial moments (i.e. at the turn of centuries, and all the more so millennia) of our history occurring changes seem to be, however, more frequent and important for us. As Bodenstedt (1993) has observed "Industrial societies do not only live with change going on, they do not only admit change, but make it a precondition of future-oriented activity. [...] Agriculture and rural areas constitute a case in question. The so-called MANSHOLT plan for agrarian structural change demonstrates how public awareness is dealing with the phenomenon of change: industrial societies turned the interpretation of agrarian structural changes into a planned process of such change.

Yet today, as Bauman(1999) notes "... changes have lost all of a sudden their "cumulativeness" and "directive tendency" assumed in advance and proved earlier or accepted without proof. [...] Perception of non-cumulativeness and non-directive tendency of changes on global scale was facilitated considerably by changes in the way of experiencing one's own world having their origin in living "at one's own home." Life rooted in the context of "late-modern" or "post-modern" world is subject to fragmentation, splitting into episodes not necessarily overlapping one another and frequently also free of interrelationships, or independent of one another. [...] Distinguishing between important changes and those of secondary importance, decisive and minor, as well as progressive and regressive changes on global, micro-social or individual scale loses its epistemological roots, which used to be saddled in experiencing personal life as "realisation of a project"."

The above diagnosis makes also sense in relation to the substantive stratum of debates. A young man entering the labour market today has to reckon with changing their qualifications several times. The place of dwelling is not a mitigating circumstance; it concerns, to the smallest degree, inhabitants of rural areas.

The time issue is linked with the space issue by the issue of a changing place held by agriculture in the post-industrial era. It is noted in this context that it loses the character of a distinct and separate sector melting into a diversified national and rural economy (Basile, Cecchi, (2000). Although as Williams(1989) has stressed, the "rural economy has never been solely an agricultural economy only as a consequence of the Industrial revolution [has] the idea (though never the full practice) of rural economy and society been limited to agriculture""(seeWhatmore,1994). Consequently, in most European countries agriculture has come to play a marginal role in national economy, in the process of value added and employment creation, whereas agricultural producers cease to be the main social stratum in villages (Hunek, 2000). This process has been described extensively in the literature of the subject. Shucksmith and Chapman(1988) have noticed, that "In terms of market and economic forces, the declining importance of agriculture and other primary activities and the growth of the service sector is well known. Many rural areas are now growing faster than urban districts, while others experience decline: the economic and social processes underlying this diverse trends are not always well understood. One key element is the increasingly global penetration of local markets, with many rural areas and firms seeking to protect themselves from global competition by creating local products which depend on local identity for their market niche. There is a general shift to a service-based economy in which the information and knowledge-based industries play an increasing role, bringing both opportunities and threats to rural areas."

The term 'disagrarisation' appearing in this context is sometimes understood both as a process of declining (but all the time significant) importance of agriculture within the national economy (see: Hunek, 2000,) and as transformation of the European space (Kuklinski, 2000). Kuklinski perceives transformation issues in the European space in the context of integration, globalisation and metropolisation processes. Mechanisms accompanying these processes are phenomena of durability, decay and metropolisation. Making reference to results of the IST Conference (1998), the European space is treated as a dynamic mosaic of agrarian, industrial and informative spaces with emphasis laid on decay processes of agrarian and industrial spaces (for instance, process of eliminating coal mining from the landscape of Western Europe) and as a process of dynamic development of the information society space. Simultaneously, it underlines complexity and sluggishness of the transformation process of agrarian space in the EU countries, which are due to the leverage of pressure groups striving to preserve the Common Agricultural Policy in an unchanged form. However "The 20th century has to bring an appropriate solution of this problem, among other things, through weakening production functions of the agrarian space and strengthening the role played by this space in shaping the European ecological rural landscapes." (Kuklinski, 2000).

The above dilemmas have become an important element in the discourse about European village and agriculture, which has intensified, in particular, during the last decade. As Ray (2000) observed "Across Europe, and indeed "advanced industrial" countries in general, the welfare state model is being incrementally transformed. The privatisation of state utilities, the commissioning of agencies to deliver services under contract to the state, the new emphasis on sub-state entities (that is, local administrations, communities and individuals) to take responsibility for their own well-being, and the consequent requirements to devise new modes of management by the "centre" are the characteristics of the emerging ethos (Rose 1995; Palfrey and Thomas 1996; May and Buck 1998; Marsden and Murdock 1998). The new political ethos is variously portrayed as a necessary response to the pervasiveness of liberal-democratic political institutions and market-oriented economics of the new world order (for example, Fukujama 1995), or as a mass delusion by the "myth of globalisation" leading states to bow to the primacy of the market and to relinquish collective gains (for example, Bourdieu 1998).

Initiated in the academic milieu it has become also a particularly important element in political activities in recent years. European policy will now be re-oriented; it will increasingly become multisectoral. According to Djurfeldt (1999) "The new buzzwords in official discussions of agricultural and rural policy are **"multifunctionality"** and **"multisectorality"**. The first term refers to the fact that agriculture has several functions, not only to produce food as such, but also to offer for sale special localised products, provide a beautiful landscape, fresh environment, biological and cultural diversity, etc. The new direction in EU policies is to support the multifunctionality of agriculture - a thrust that begun already with the Agenda 2000, initiated a few years ago." During the debate which took place during the XVIII Congress of the European Society for Rural Sociology, Van Depoele (1999) has noticed, that the Union is now going one step further, by subordinating its agricultural policy to its rural one. The latter is *multi-sectoral*. It recognises the fact that agriculture accounts for a diminishing share of rural development. At present 5.3 per cent of the European labour force is engaged in agriculture, and the sector is losing 200 000 jobs a year.

Agriculture is no longer the backbone of rural areas. That's why a policy of subsidising farming, like the CAP, cannot prevent the desertification of the countryside" (see: Djurfeldt, 1999).

It is difficult, however, to leave this prediction without any comments. The first of them derives from the latest history of modernisation of Western European agriculture. On the basis of development of the French agriculture, its intensification and main effects of its evolution Lamarche(1998) formulates a view – shared not only by him – that a productivistic mode of development has already reached its limits and has to be changed. It cannot be continued both due to problems, which it generates (i.e. economic, social and ecological problems), and because of a very rapidly changing global socio-economic context. "Goals of agriculture development policy cannot be the same all the time. [...] All people responsible for agriculture (both politicians and farmers themselves) are aware of it today, but instead of contesting fully the productivistic model they prefer to make efforts aiming at introduction of dual agriculture. The first model, i.e. the productivistic model "is reserved for the elite of more and more efficient producers," whereas the main goal of the second model is sale of services in such fields as development of agro-tourism, care and concern about landscape, communal services, and so on. "Development of this model aims at stabilisation of these rural and farming populations, which have remained outside the main trend of productivistic development or they already pushed towards the margin" (Lamarche, 1998). According to the scholar quoted here, such activity cannot solve the crisis definitely. At best it can prevent dramatic effects of a progressive process of social marginalisation. According to Djurfeld (1999) "We here seem to get a scenario for the future of European agriculture.[...]On the one hand, we have the commercially oriented farmers, entrepreneurs operating entirely on the terms set by the market, and intent to survive by making profit from production. On the other hand we have the rest, who for various reasons are unable or unwilling to produce on commercial terms. [...] While the commercial farmers produce the real thing, i.e. the food, the remaining eighty per cent of farmers will be constrained to producing the landscape, the sustainable environment, and the touristic attractions that Europe's city-dwellers demand in their holidays. Is this a sustainable model of agricultural and rural development? And is it one we want?"

The issue of social exclusion has become in this way one of the most important issues not only for rural areas (see, for example: Shucksmith and Chapman, 1998) but for the entire society as well. According to Dahrendorf (1997), among the greatest contemporary threats faced by the European societies is the necessity of looking for a new equilibrium between prosperity, social solidarity and personal freedom. Competition, the stake of which more and more frequently is "success and wealth" or "defeat and poverty", destroys social ties, solidarity, co-operation. "Competitiveness has become a value pushing out solidarity and social ties, owing to which the protective states were able to operate. The global competition of the 1990s - between economic systems, states, large corporations, but also between all of us and each one of us separately - has shaken the system of values and the entire model of Europe's life. Our economy can withstand this competition, but the whole society pays for its success(...) Maybe we simply go through a process of radical transformations resembling the period of an extremely painful transformation of the feudal societies into the capitalist ones. We do not know where this process leads. [...] The next decade will bring phenomena, in the face of which it will not be easy to defend both prosperity and freedom. And it will be very difficult for us to defend the 19th century canon of our European values: freedom - solidarity - prosperity. In the world of global competition it will be threatened - to an increasingly greater degree - by the Asian values (...) and primarily by very high productivity and social solidarity forced out by the State"

In this presentation an attempt will be made to characterise chosen Central and Eastern European countries, with a special emphasis placed on Poland[1]. It is the author's ambition to accomplish this goal not only in the descriptive layer (that is, sociographic) but also combine it with an attempt to determine conditions, states and mechanisms of attitudes towards selected aspects of the present time and life plans. It will be supplemented by presentation of the most important assumptions of the Cohesive Structural Policy of Development of Rural Areas and Agriculture in Poland (the Government programme from 1999, which is to be implemented by means of a number of operational programmes including the SAPARD Programme - *Special Accession Programme for Agriculture and Rural Development*).

We would like to open the next part of our presentation with a statement about trends visible in Central and Eastern European countries, which orient the main structures of agriculture "... towards models typical of market economies. It means that the convergence process of agriculture in Central and Eastern European countries with agriculture in the European Union countries has been already started and in some cases a considerable progress has been made in this field. It seems that the process of convergence and harmonisation of agriculture in CEECs with its development models in Western Europe will be a fundamental factor determining behaviour of agriculture in Poland in the early years of the 21st century" (Hunek, 2000). It is difficult, however, to leave this prediction without any comments. The first of them derives from the latest history of modernisation of Western European agriculture.

Methodology

Rural development is a multidisciplinary topic. The literature in the field of social sciences dealing with rural development concepts makes also allowances for both economic, demographic, environmental and health issues and for strictly sociological issues including cultural processes and variations[2]. The term 'rural development' uses the category of rurality, which is a category of interrelated social-cultural and economic spheres. In line with the UNDP social development concept, its basic objective should be the broadening of possibilities of making different choices by people so that transformation processes can be characterised by a greater involvement of those interested and have a more democratic character. For rural development issue the methodological directive of "putting people first" (Cernea, 1991, Kottak,1991, Uphoff,1991, and others): ensues also from understanding the fact that changing economic characteristics of rural areas imply also that attitudes undergo major changes (that is, cognitive, axiological and behavioural components), or – as we conventionally say – mentality of rural communities.

Within this approach there can be also found two Polish authors (let us underline that both are economists), who analyse problems of development of rural areas in Poland in the light of future membership in the European Union. Identifying barriers and priorities from the viewpoint of Poland's integration with the European Union they underline that – without negating the deficit of economic infrastructure and funds – the most important barriers to rural areas development are inherent in people.(Klodzinski,Wilkin,1998).

The change of political and economic system started in 1989 was accompanied by numerous transformations also in the village and agriculture. Meanwhile, the situation of peasants or farmers and the situation of rural population in the period preceding systemic changes has been generating different assessments and interpretations both among researchers, politicians, and those directly interested including the remaining Poland's population till the present day. Hence, defining the very essence of occurring processes and their effects tends to vary significantly in this case according to a subjectively defined situation. Therefore, the Weberian programme of explaining social systems through purposeful actions of individuals, characteristics of structures and processes resulting from deliberate actions of individual actors perceived, however, within the framework of market institutions, formal organisations, etc. would be quite useful here. The present paper aims at an analysis of selected issues referring primarily to the village and agriculture in the situation of systemic transformation taking place in Poland and other countries of this region nowadays. The author shares the view of Skąpska (1995) and Ziółkowski (1998) that transformation "is a conscious effort of introducing a new global 'institutional system" [...] made, however, in conditions determined by the inherited level of socio-economic development and the existence of traditional, habitual ways of perceiving and operating in social reality. Recognising that these transformations can be treated as transformations of old and creation of new institutions it should be underlined that both the formation process of new institutions understood as organised and distinguished types of activity (for instance, commodity exchanges) and also changes in institutions themselves "...activity of individuals, social categories and groups changes in the broadest sociological meaning, there appear new models of behaviour, thinking and valuation, new adaptation strategies and new ways of coping with reality. These changes have a different rhythm and are a resultant of central decisions, heritage of the past and mass, spontaneous processes".

Results

The first goal of this part of the paper is to verify a hypothesis about progressive individualisation of social life in local rural systems and to consider perspectives of collective actions. We are accompanied by an assumption that "Local space plays, to an increasingly big degree, only the role of a prerequisite for mutual associating of inhabitants rather than that of a determinant of their social life" (Starosta,1995). The basis for verification was made changing forms of mutual assistance among population of village and small town local systems.

The empirical data presented in this part of paper were gathered within an international research project entitled "Patterns of Social Participation and Social Structure of Local Communities in Bulgaria, Poland and Russia" funded by the Scientific Research Committee of the Polish Academy of Sciences, (grant 1 HO 1F 016 09) and partly by Universite du Quebec a Rimouski (Canada). Altogether the surveys encompassed 3 Polish communities (Poddebice, Dobron and Siemkowice) representing the Administrative Province of Sieradz: central part of Poland, 2 Russian communities (Ilinskoye and Luch) representating the Administrative Province of Ivanovo (central part of Russia), 3 Bulgarian communities (Gavrilovo, Biala and Kotel) representing the Kotel District in central Bulgaria. All in all, 1616 questionnaires were collected containing the same set of questions in each country.

One of the main methodological assumptions was choosing in each country the communities consisting of 1,000 to 100,000 inhabitants differentiated with regard to economic, political and social activity. In each country (with the exception of Russia where only two localities were surveyed) there were chosen for the surveys: one community with under 1,500 inhabitants, one with the number of inhabitants between 1,500 and 5,000 and one with over 5,000 inhabitants. The respondents were random chosen for questionnaire interviews from the registers of voters in particular localities. The collected data fulfil the requirement of representativeness only for the communities in the survey, but they are not representative for particular countries. The collecting of data was completed in February 1997.

We would like to start the presentation of survey findings by showing the general level of satisfaction with the place of residence in the communities under survey.

Table 1 Levels of satisfaction with rural and small town communities by country

Levels of satisfaction	BULGARIA		POLAND		RUSSIA		TOTAL	
	N	%	N	%	N	%	N	%
1.Very satisfied	64	10.5	140	22.5	24	6.0	228	14.1
2.Rather satisfied	170	28.4	327	38.0	258	64.5	755	46.7
3.Ambivalent	218	36.4	106	17.2	80	20.0	404	25.0
4.Rather dissatisfied	112	18.7	37	6.0	32	8.0	181	11.2
5. Dissatisfied	35	5.8	7	1.1	6	1.5	48	3.0
Total	599	100.0	617	100.0	400	100.0	1616	100.0

Source: (Starosta, Rokicka, 2000)

As Starosta and Rokicka(2000) have noticed "The data included in Table 1 indicate that about 61% of all the respondents declared positive attitudes towards their place of residence. Ambivalent attitudes were expressed by 25%, and negative ones by only 14% of the respondents. The variations in emotional attitudes between particular countries are quite interesting. The inhabitants of Russian villages and small towns rank first with regard to positive attitudes (70% - satisfied, 20% - ambivalent attitudes, and 9% dissatisfied) and followed by the inhabitants of Polish communities (60% - satisfied, 17% - ambivalent attitudes, 7% - dissatisfied). The most negative opinions were declared by the inhabitants of Bulgarian communities (39% - satisfied, 36% - ambivalent attitudes, and as many as 24% - dissatisfied).

Satisfaction with the local environment is generally quite high in the surveyed countries with the exception of Bulgaria. It could be added that the share of rural and small town inhabitants satisfied with the local environment in Poland has not changed for over ten years and it ranges between 60% and 70% (Starosta,1995) Thus, the systemic change here has not exerted any major influence.

In order to determine the role of the natural environment as a determinant factor of positive or negative attitudes towards the place of residence we were asking about the respondents' satisfaction with different fields of their community life.", see Table 2 overleaf.

Following Starosta and Rokicka(2000) "The results of this analysis allow us to draw the following conclusions:

The natural environment as a determinant factor in the evaluation of local life quality is rather insignificance in rural communities of Poland Russia and Bulgaria. The local satisfaction results mostly from the good quality of family life, good social relations with other residents in the community in all the localities under survey. [...] Residents of the communities under survey were least satisfied with the activities of the political parties, their present income and local job opportunities. The last factor of dissatisfaction is a relatively new phenomenon in Eastern European Countries since the Second World War. It appeared in the nineties as a result of both the transformation from a planned to a market economy and as an inclusion of the former socialist societies into the global world economic order." (Starosta and Rokicka, 2000)

Table. 2 Area of satisfaction in community life

Satisfaction with:	Percentage of satisfaction with the place of residence and the ranking					
	Bulgaria		Poland		Russia	
1. Family relationship	88.8	1	86.4	1	82.3	1
2. Social relations in place of residence	53.8	3	70.6	5	62.1	2
3. Flat/ house	84.4	2	82.5	2	58.5	3
4. Present incomes	9.9	14	23.4	13	12.5	12
5. Job opportunities in vicinity	12.5	13	21.4	14	27.4	8
6. Health service	36.8	8	55.3	10	32.6	7
7. Rest and recreation conditions	28.6	9	31.1	12	23.2	9
8. Quality of natural environment	17.0	12	43.3	11	9.9	13
9. Transport services	38.9	7	80.8	3	33.2	6
10. Quality of services for population	23.9	10	72.0	4	9.1	14
11. Local public administration	44.7	6	66.4	8	16.5	11
12. Activity of schools and educational institutions	52.0	4	69.5	6	56.2	4
13. Activity of local church	23.1	11	68.0	7	42.9	5
14. Mayor's activities	48.6	5	65.2	9	20.4	10
15. Political parties' activities	6.8	15	8.6	15	2.9	15
N	599		616		400	

Source: Starosta, Rokicka, 2000

It is commonly recognised in the sociological literature that co-operation between producers in the process of agricultural production is deeply rooted in rural local communities. It would, however, be certainly more appropriate to use the past tense in this case. Paraphrasing Sennet we can say that a remarkable characteristic of co-operation issue are frequent references made to the past. Considering an issue of the ways inhabitants of local communities take part in more or less formalised kinds of economic co-operation, we have analysed their knowledge and their rate of participation in community actions aimed at solving problems, which were important for local environment. The subject of our research included also various kinds of informal help (mutual or asymmetrical), by which we meant only the broadly understood sphere of economic activity of the household (i.e. help rendered others or received from other persons - households in the form of work, material goods (borrowed or granted) or money. We were interested how in those three differing (from the point of view of economic and cultural or historical background, including different experiences with communistic system) European countries evolve the forms of social participation in the local communities. We have limited our analysis to informal economic relationships between households. We have analysed their behaviours and opinions regarding the scale of above mentioned processes in a given community, both in the past and at the present moment. Searching for the precise indicators of the nature of processes under consideration and the ways they differed from one another, we have analysed statistical explaining capacity of such an independent variables as gender, level of education, main source of income, self-estimation of personal wealth, social status and, finally, the character of the living place (house, rented or owned apartment). In the process of explaining of 22 dependent variables using 6 dependent variables it turned out, that the most frequently and efficiently (using Cramer's V measure of the strength of the relationship) differing is the variable describing the country of origin of the researched subjects (it explained statistically important differences for 17 dependent variables). The next five independent variables (the level of education etc.) explained at statistically significant level similar number of dependent variables (9 - 12). We found out that gender was the factor having the least influence on behaviours and opinions. Having considered the main purpose of the research and above mentioned results the further analysis was concentrated on mainly on showing the forms of mutual help which the members of local communities give to each other in three researched countries. It is to be stressed that preliminary statistical analysis conducted separately in relation to each of the countries hasn't found any significant differences in the ways of rendering mutual help in each of the communities. The country of origin remained the most efficiently differing factor in relation to behaviours and opinions. More evidence supporting those statements is presented below in Table 3.

Table 3. Frequency of informal (mutual) help, that takes place in the local environment in relation to economic activities undertaken with regard to house, apartment or something else (as it is reflected in the opinions of inhabitants)

Country	Yes, always	Yes, often	Yes, from time to time	No	Others
	%	%	%	%	%
Poland	9.2	17.9	46.8	15.9	10.2
Russia	2.2	36.1	35.6	10.4	15.7
Bulgaria	11.4	22.9	53.4	2.0	10.3
Total (%)	8.3	23.9	46.5	9.4	11.9
N	134	386	750	151	191

[*** Cramer's V]

There are differences however in the believed intensity of those activities in each of the researched countries, see Table 4 overleaf and such opinion s are most frequent in Russia, while in Poland they differ the most. Farmers and pensioners more often recognise existence and significance of those processes than other segments of the population.

More than ten years ago members of local communities helped each other more frequently. The hypothesis assuming that the phenomenon of mutual help disappears was proven correct. It is supported by the opinions of those respondents, who believe, that people were more helpful in the past (54%). It is to be emphasised however, that significant part of the respondents doesn't recognise any changes in this area (24%); in Poland one of every three respondents is unable to realise that difference.

Table 4. Dynamics of the phenomenon of mutual help in the local environment (as it is reflected in the opinions of inhabitants).

Country	more frequently in the past	no changes	more frequently today	hard to say
	%	%	%	%
Poland	39.3	29.2	2.2	29.4
Russia	41.6	35.5	3.3	19.6
Bulgaria	77.9	11.4	1.2	9.5
Total (%)	54.3	24.1	2.1	19.7
N	867	385	33	312

[*** Cramer's V]

In the author's opinion, the research findings presented above verify positively the hypothesis advanced earlier about a progressive rationalisation process of social life (including also a formal rationalisation process) in local communities (that is, village and small town communities) in Poland. Simultaneously, the results of above comparative analysis seem to enable a positive verification of a thesis about a much smaller progress made in this process in the analysed local communities in Russia and Bulgaria (where traditional forms of mutual assistance are more common).

It seems that the research findings presented above make it possible to verify the hypotheses formulated earlier pointing to different forms of collapse of traditional mutual assistance forms according to scale and pace of processes of systemic transformation, which, simultaneously, inscribes itself into more general globalisation processes. Moreover, it seems that the conclusions flowing from the analysis make it possible to verify positively a hypothesis about a progressive individualisation of social life (and, thus, broadening the sphere of personal freedom also in local communities of the Central and Eastern European countries showing also variations in the character and pace of this process in Poland, Russia and Bulgaria).

The second goal of this presentation is to describe some changes in the peasant system of values.

Analyses of the place held by land in the peasant system of values stress its central and autotelic place. Styk(1999) includes land in the fundamental axiological triad underlines that: it constitutes a foundation of each agrocentric socio-cultural system as the main component and basis of an agricultural farm. The analysis below will be focussed on the process of changing values of farmers in relation to two key and, simultaneously, interconnected elements of their *micro-cosmos*, that is, land and farm. It will be done on the basis of data coming from surveys carried out in: 1988 (sample of 233 farmers), 1990 (140), 1993 (211), and 1995 (160)(Pilichowski,1999). Researchers frequently point to the accuracy of using the perspective of land and farm for describing the process of transformations. Let us remember here that whereas at the foundations of the system of values in traditional societies lay a view that land should be absolutely possessed to be able to live and work, today the value of land and work has been altered in a fundamental way. The directions of these changes and their socio-economic explanations will be presented below. In a synthetic commentary the author wishes to stress:
- continuity and change of particular indices characterising normative dimensions of farming;

- obtaining through the factor analysis two dimensions describing farmers' subjective convictions about the peasant micro-cosmos. The first dimension has been called an axiological perspective, and the second - according to the author - is described well by the Weberian formula of formal rationality.

- the analysis of quantitative distributions reveals a minority group of farmers sharing the traditional peasant axiology and a majority group of farmers largely questioning traditional values.

- the performed explanatory analysis using a number of socio-economic variables showed a weak grounding of differentiated images about farming in socio-demographic variables and much more significant in variables of economic situation and awareness state.

- the built regression model explaining in greater detail variations in axiological perspectives points to a system of variables explaining these issues most fully; (all independent variables included in the tables are significant at the minimum level p=0.1).

An important symptom of a changing normative order in the farming community is a change in attitude to land. It loses its sole utilitarian value and becomes both a commodity, in which capital can be simply invested, and also a factor, which can be exploited ruthlessly for maximisation purposes. Thus, a hypothesis about consolidation of a conviction in many farmers' awareness that a larger farm is not only a better farm but it is also indispensable to survive in the market has been confirmed in the light of results yielded by this analysis. It gives rise to the conception of a changed role of the family farm owner, who changes from a farm owner to an agricultural producer. It releases also a mechanism of consolidation and polarisation of the agrarian structure, as a result of which a considerable part of farmers become eliminated from the market and quite a big part of them resign also from farming (from among the extremely rich literature of the subject a reference could be made in this place to the French experience described by: Mendras, 1970; Lemarche, 1992; Ramboud, 1994; and Broussard, Delorme, Servolin, 1998).

It was the author's intention to show changing patterns of farming production in the Polish agriculture in three fields distinguished by Sztompka (1995) and make an attempt to show patterns, schemes, and rules concerning farmers' activity and thinking.

In accordance with this typology there was performed an empirical analysis, in which an attempt was made to include both the behavioural level (what people do) and the psychological one (what people think, know). The deepest level inferable both from behaviours and views, that is, patterns, schemes, rules, codes concerning farmers' activity and thinking should, thus, emerge from a further analysis of the categories presented above. However, this is not an object of the present paper. Hence, an object of closer analysis were made farmers' subjective convictions concerning key elements of *the axiological triad of peasant micro-cosmos: land, family and work* (Styk, 1999:7). Getting to know them better was to facilitate a more comprehensive description and become an important element in characterising the world rationalisation process (on the example of rural farming reality).

In these studies the author was interested primarily not so much in individual events or opinions but rather in trends of perceiving, evaluating, experiencing and responding to the social reality. It was, first of all, at this level that the main part of analysis was performed, which allowed to distinguish different dimensions of axiological orientation and formal rationality (as it is expressed by Bauman, 1996: *the voice of heart and moral obligations, and the voice of mind, sphere of interests - profit*), scale of psychic discomfort, openness to and readiness to embark upon change, scale of deprivation or, finally, dimension of subjectivity - orientation at claims.

In this light, *the way of outlining a portrait* of people involved in farming from the viewpoint of rationality presented by them assumes an extremely big importance. These issues play an important role in the analyses presented here. Reference was made in them, among other things, to typologies of Halamska(1996) and Gorlach(1995). Gorlach (1995) (conception of four types of farmers' rationality) distinguished on the basis of four variables (that is, a question about machines and equipment for agricultural production possessed by a farm, a question about the way of defining the farm owner's role, a question about propensity to choose again the farmer's professional role, and a question about opinions on usefulness of a lease as a way of enlarging a peasant farm's area) a peasant, a post-peasant, a pre-farmer and a farmer rationality types.

On the other hand, if a reference was made in this place to Maurel and Halamska(1996)'s (concept of four main groups of agrarian rationalities (characterised by land trajectories of a farm, that is, its enlargement, stabilisation or reduction, and attitude to land measured by purchase-sale intentions), i.e, *resigned, frustrated, satisfied and entrepreneurial farmers*, it would be necessary to confirm their empirical presence also in the empirical material presented here.

In the light of the author's experience, the most important issue appeared to be not that of typology but rather of variables - indices used to describe the occurring process of changing rationalisations and the type - way of their socio-economic grounding. With regard to the social rationality issue it would be worthwhile to quote an opinion of J. Styk(1993), who is convinced that *these rationalities converge at some point to create a non-adjectival rationality and, thus, also an economic and social rationality.*

Final remarks

In this paper we have analysed three important sociological issues or barriers of rural development, mainly in Poland. In Poland we can include, first of all, low education level of rural population: the share of people with higher education is five times smaller than in towns (i.e. 2%), and that of people with secondary school education is twice smaller (i.e. 14%). An offer of rural educational units is, as a rule, more modest than in urban units. An opportunity for overcoming these variations is a formula: "Internet in every school," and development of permanent education within the Distance Learning System. Four percent of primary school graduates in rural areas do not continue their education. Meanwhile, the type of school chosen by remaining graduates is usually determined by financial considerations. Consequently, most primary school graduates choose schools located in the vicinity of their villages. The share of rural youth with higher education is much smaller in comparison with youth in cities. It results in a low level of educational aspirations. As has noticed Szafraniec 2000)"There is no doubt [...] that there is one more factor putting rural youth in an even worse position and confronting them with a much more difficult task than their town counterparts. Rural youth are poorly educated, the majority are deprived of roots, they have meagre chances of achieving their goals and fulfilling aspirations (economic and cultural poverty of rural families, the poverty of rural education), moreover, they are in danger of being unemployed."

The second barrier is the level of involvement in community activities. Low involvement in community actions, small scale of informal collaboration are a good confirmation of decaying traditional social bonds in local communities in CEECs. This issue was discussed in greater detail in this paper.

The third barrier, but also a potential development opportunity, is perceived in the state of awareness (mental construct) of rural population. Surveys carried in the 1990s provide a number of arguments allowing to formulate a hypothesis about a considerable potential inclination of not a small part of farmers to abandon farming. However, an alternative labour market is indispensable, which is an economic and not necessarily agricultural development of rural areas. Surveys conducted in 1999 by the Institute of Public Affairs show that over fifty per cent of farmers would resign from running a farm in favourable conditions and only 13 % would like to see their children in the role of farmers in the future (Instytut Spraw Publicznych,1999,).

It is underlined in the quoted report about Rural Areas Development that "The Polish village apart from economic infrastructure or funds needs primarily knowledge, motivations, inclinations and skills of interacting and autonomy."(Klodzinski and Wilkin,1998). Guided by this diagnosis and results of above analyses, a postulate concerning the necessity of activities aimed at releasing initiative and active involvement of rural inhabitants becomes justified. Such opportunities are created by, for instance, the LEADER Programme implemented in the EU. The past experience connected with this programme indicates that domestic and EU funds can contribute to considerable mobilisation of funds from other sources (local, private).

The LAG formula (*Local Active Group*) proves, on the other hand, to be a good way of local mobilisation (...) and, according Kovach(2000), an opportunity for restricting the impact of bureaucratic administrative structures The answer to Christopher Ray's question about usefulness and implementation of the LEADER Programme in CEECs is, thus, positive[3].

It was the author's wish and also goal to contribute to a better insight into, understanding and in this way explanation of this social peculiarity sphere, which is directly connected with farming economic activity at the time of Great Transformation in Central and Eastern Europe.

The above diagnosis is also, at least partly, relevant to Poland and other countries in this part of Europe. We live at the time of deep social changes. Although in Poland (but also, for instance, in the Czech Republic or in Hungary) "The systemic transformation understood as transition to capitalism and democracy ("democratic capitalism") has been crowned with success" (Ziolkowski, 1998) "... the analysis of present time as *sui generis* reality" performed in a new context of "... dynamics, diversification and internal tensions in the world capitalist system (Ziolkowski, 1998) (with the place held by these countries in this system being neither its core not absolute peripheries) becomes more appropriate for analysing the process of changes taking place in these countries than the transformation perspective ("from or to"), the fact of transition to democracy and capitalism does not mean that a historical process of changes has been finished. It becomes, on the other hand, "a causative agent, which makes possible or, at least, facilitates quite a deep transformation in the entire Polish reality – not only in politics and economy but also in social life and culture"(Ziolkowski, 1998).

References

Basile,E., Cecchi,C.,2000. Formowanie sie gospodarki obszarow wiejskich - analiza ekonomiczna (Formation of Economy in Rural Areas – Economic Analysis). In: Hunek, T., ed., Dylematy polityki rolnej. Integracja polskiej wsi i rolnictwa z UE. (Dilemmas of agricultural policy. Integration of Poland's countryside and agriculture with the European Union). The Foundation for Assistance Programmes to Agriculture (FAPA), Warszawa.

Bauman, Z., 1999. Po co komu teoria zmiany? In Kurczewska, J., ed., Zmiana spoleczna. Teoria i doswiadczenia polskie. IFiS PAN, Warszawa.

Bauman,Z., 1990. Thinking Sociologically. (Polish edition 1996 by Zysk i S-ka, Poznan).

Beck, U., Giddens, A., Lash, S. 1994. Reflexive Modernization. California, Stanford: Stanford University Press.

Bodenstedt, A., 1993. Przyszłość wiejska: wyzwanie dla naszego sposobu myślenia (Rural future: a challenge to our way of thinking) In Wies i Rolnictwo, 3 (80), IRWiR PAN, Warszawa.

Broussard, J-M., Delorme, H., Servolin, C. 1998. Rolnictwo nowoczesne - doświadczenia francuskie, w: Z. T. Wierzbicki, A. Kaleta (red.), Rolnictwo i wieś europejska. Od korzeni ku wspólnej przyszłości w XXI wieku. Toruń-Warszawa: Uniwersytet Mikołaja Kopernika, p. 195-209.

Cernea, M.M., 1991. ed., Putting people first. Sociological variables in rural development. Second edition. Revised and expanded. Published for the World Bank Oxford University Press.

Dahrendorf,R.,1997, Nie o takiej snilismy Europie (w rozmowie z Jackiem Zakowski) In Gazeta Wyborcza, 1-2 July, Warszawa.

Djurfeldt, G., 1999. From CAP to CAR? New directions in European rural policy. In XVIII Congress Of the European Society for Rural Sociology, Lund , Sweden, 24-28 August 1999. How to be Rural in Late Modernity. Report and Evaluation, Lund University).

Giddens, A., 1998. The third way. Polity Press, Cambridge.

Gorlach, K. 1995. Chłopi, rolnicy, przedsiębiorcy: "kłopotliwa klasa" w Polsce postkomunistycznej. Kraków: Uniwersytet Jagielloński.

Halamska, M. 1992. Gospodarstwo rodzinne w Polsce (Family Farm in Poland).
w: H. Lamarche (ed), Rolnictwo rodzinne. Międzynarodowe studium porównawcze, Część I Rzeczywistość polimorficzna, Warszawa: IRWiR PAN.

Halamska, M. 1996. Polskie gospodarstwo rolne z perspektywy porównawczej,
w: M. Kozakiewicz, (red.), Wieś i rolnictwo w badaniach społeczno - ekonomicznych. Warszawa: IRWiR PAN.

Hunek, T., 2000. ed., Dylematy polityki rolnej. Integracja polskiej wsi i rolnictwa z UE. (Dilemmas of agricultural policy. Integration of Poland's countryside and agriculture with the European Union). The Foundation for Assistance Programmes to Agriculture (FAPA), Warszawa.

Hunek, 2000, Uwarunkowania i mechanizmy rozwoju gospodarki obszarow wiejskich w Polsce (Determinants and Mechanisms of Development of Rural Areas Economy). In Hunek, T. ed., Dylematy polityki rolnej. Integracja polskiej wsi i rolnictwa z UE. (Dilemmas of agricultural policy. Integration of Poland's countryside and agriculture with the European Union). The Foundation for Assistance Programmes to Agriculture (FAPA), Warszawa.

Instytut Spraw Publicznych (Institute Of Public Affairs), 1999. Rolnicy i mieszkańcy wsi o przyszłości polskiej wsi i rolnictwa w obliczu integracji z Unia Europejska, Warszawa.

IST "98 Vienna, Living and Working in the Information Society, European Commision, IST Conference, Vienna, 30 November - 2 December 1998.

Kaleta, A., 1999. Revitalization of rural areas. In Starosta, P., Kovach, I., Gorlach,K., 1999. eds, Rural societies under communism and beyond. Hungarian and Polish perspectives. Lodz University Press, Lodz.

Klodzinski, M., Wilkin, J., 1998. Rozwoj terenow wiejskich w Polsce w swietle przyszlego czlonkostwa w Unii Europejskiej (Development of Poland's rural areas in light of its future membership of the European Union). In Wies i Rolnictwo, 4 (101), IRWiR PAN, Warszawa.

Kovach, I., 2000. LEADER, a new social order, and the Central- and East-European Countries. Sociologia Ruralis, 40 (2): 181-189.

Kuklinski, A., 2000. Transformacja przestrzeni europejskiej. Artykuł dyskusyjny (Transformation of European Space. Controversial article). In Studia Regionalne i Lokalne,1(1), pp.25-32.

Lamarche, H., 1998. Francuski model rolnictwa czyli paradoksy udanego rozwoju. (French Model of Agriculture or Paradoxes of Successful Development. Agriculture and European Village). In: Wierzbicki, Z.T., Kaleta, A., eds, Rolnictwo i wies europejska. Od korzeni ku wspolnej przyszlosci w XXI wieku (From Roots to Common Future in the 21st Century).UMK, IRWiR PAN, Torun - Warszawa,pp.201-209.

Pilichowski, A.,1999. Land in the Polish agrarian system, In Starosta, P., Kovach, I., Gorlach,K., 1999. eds, Rural societies under communism and beyond. Hungarian and Polish perspectives. Lodz University Press, Lodz.

Pilichowski,A., Stolbov,W., 2000. Nieformalnaja kooperacja w selskich obszczinach, In Sociologiczieskije Issledowanja, ISSN 0132-1625, Moskwa.

Ploeg, J.D. van der, Long, A., 1994. Endogenous development: practices and perspectives. In Ploeg, J.D. van der, Long, A., eds, Born from within. Practice and perspectives of endogenous rural development, Van Gorcum, Assen, pp.1-6.

Ploeg, J.D. van der, Saccomandi, S., 1995. On the impact of endogenous development in agriculture. In Ploeg, J.D. van der, Dijk, G. van, eds, Beyond modernization, Van Gorcum, Assen, pp.10 27.

Putnam, R., 1993. Making democracy work. Civic traditions in modern Italy. Princetown University Press.

Ray, C., 2000. The EU LEADER programme: rural development laboratory. Sociologia Ruralis, 40 (2): 163-171.

Shucksmith, M. and P. Chapman, 1998. Rural development and social exlusion. Sociologia Ruralis, 38 (2): 225-242.

Shucksmith, M., 2000. Endogenous development, social capital and social inclusion: perspectives from LEADER in the UK: 208-218.

Skąpska, G. 1995. Reforma ekonomiczna i kształtowanie się porządku instytucjonalnego w Polsce. Dziedzictwo przeszłości, mity i wizje przyszłości, w: A. Sułek, J. Styk (red.), Ludzie i instytucje. stawanie się ładu społecznego. Lublin: Wydawnictwo Uniwersytetu Marii Curie-Skłodowskiej, t.1, s.139-154.

Spojna Polityka Strukturalna Rozwoju Obszarow Wiejskich i Rolnictwa (Cohesive Structural Policy of Rural Areas and Agriculture Development), 1999.http://www.minrol.gov.pl

Starosta, P., 1995. Poza metropolia, Lodz University Press, Lodz.

Starosta, P., Draganowa, M., 1999. Social identification with local communities and the globalization process in rural areas of Eastern Europe, In Kasimis,CH., Papadopulos, A.G.,eds, Local responses to global Integration, Aldershot, Ashgate.

Starosta, P., Kovach, I., Gorlach,K., 1999. eds, Rural societies under communism and beyond. Hungarian and Polish perspectives. Lodz University Press, Lodz.

Starosta, P., Rokicka, E.,2000. Environment and the quality of life in the Eastern European Peripheries, In Eastern European Countryside,6' 2000, Torun, Nicolaus Copernicus University.

Styk, J. 1993. w: Wieś-Rolnictwo-Wolny Rynek. Dyskusja z prof. Leszkiem Balcerowiczem. Warszawa: Instytut Macieja Rataja, s. 35-37.

Styk, J. 1999. Chłopi i wies polska w perspektywie socjologicznej i historycznej. Lublin: Wydawnictwo Uniwersytetu M.Curie-Sklodowskiej.

Szafraniec,K.,2000. The countryside and the young rural generation. From social advancement to social marginalization. In Eastern European Countryside,6' 2000, Torun, Nicolaus Copernicus University.

Sztompka, P. 1995. Cultural and Civilizational Change: The Core of Post-communist transition, w: B. Grancelli, (ed.), Social Change and Modernization. Lessons from Eastern Europe. Berlin-New York: Walter de Gruyter, s. 235-247.

Uphoff, N., 1991. Fitting projects to people. In Cernea, M.M., 1991. ed., Putting people first. Sociological variąbles in rural development. Second edition. Revised and expanded. Published for the World Bank Oxford University Press.

Van Depoele.L. (1999) in XVIII Congress of the European Society for Rural Sociology, Lund, Sweden, 24-28 August 1999. *How to be Rural in Late Modernity*

Wesołowski, W. 1995. Niszczenie i tworzenie interesów w procesie systemowej transformacji: próba teoretycznego ujęcia. "Kultura i Społeczeństwo", nr 2: 3-25.

Whatmore, S., 1994. Global agro-food complexes and the refashioning of rural Europe In Thrift, N., Amin, A. (eds) Holding down the global, Oxford University Press.

Ziolkowski, M. 1995. Pragmatyzacja świadomości społeczeństwa polskiego,
   In: A. Sułek, J. Styk (red.), Ludzie i instytucje. Stawanie się ładu społecznego. Lublin: Wydawnictwo Uniwersytetu Marii Curie-Skłodowskiej, t.2, s.27-47.

Ziolkowski, M. 1998. O roznorodnosci teraxniejszosci. Pomiedzy tradycja, spuscizna socjalizmu, nowoczesnoscia i ponowoczesnoscia. Jn Slask - Polska - Europa. Zmieniajace się spoleczenstwo w perspektywie lokalnej i globalnej. Xiega X Ogolnopolskiego Zjazdu Socjologicznego, Katowice, Wydawnictwo Uniwersytetu Slaskiego, pp. 83-106.

ENDNOTES

1. Rural areas occupy over 90% of the country's total area, with two-thirds of these areas being involved in agricultural activity. The areas are inhabited by about 14.4 million people, who belong to 4.1 million households. The status of a village or a town is determined by administrative decisions. Making an estimate according to the European Commission's criterion (population density does not surpass 100 persons per square kilometre), it would appear that 32.8% of total Polish population lives in villages, and rural areas occupy 83% of Poland's total area. There are 870 towns and cities in Poland (including 654 small towns) and 56,803 rural localities among which: about 15% have less than 100 inhabitants; 66% from 100 to 500, 13% from 500 to 1,000, and 6% over 1,000 inhabitants. Almost 30% of Poles (and over 60% of rural inhabitants) are linked with farms or farming plots, and taking into account their connections with companies providing agricultural services their share reaches almost 40%. About 10 million people in Polish villages are linked with the farm but only 8% derive their incomes exclusively from work on their farms. For the remaining rural inhabitants an important role is played by other sources of income, that is, wages and transfers especially in the form of old age and disability pensions. Accordingly, although the Polish village is linked with the land it derives its livelihood, to a very significant degree, from non-agricultural incomes.
The village is afflicted strongly by unemployment. Registered jobless persons more and more frequently live in villages. Most of them belong to the category of long-term unemployment and they have lost the right to unemployment benefits. Additionally, it is estimated that a part of employment in agriculture has only a semblance of real employment, which causes that a large part of available work force is idle.
Moreover, rural areas tend to vary considerably with regard to development opportunities.

2. Assumptions concerning the future shape of Polish agriculture, which were formulated at the beginning of systemic transformation, assumed the necessity of creating favourable conditions for development of agricultural activity and stressed the necessity of dynamic development of rural areas (An Agricultural Strategy for Poland, 1990). Existence of a particularly strong interrelationship between processes of creating the sector of developmental farms and imparting dynamics to economic development of rural areas is underlined in the literature of the subject. According to one of authors, "In fact, it is impossible to create the sector of dynamic farms and next ensure their effective operation. Neither is a steady and sustainable development of rural areas possible without restructuring the contemporary Polish agriculture" (Hunek, 2000, p. 127).

3. Main determinants, assumptions and goals of the policy aiming at a steady transformation of agriculture and rural areas can be found in the government document entitled: *"Cohesive Structural Policy of Rural Areas and Agriculture Development"* (July 1999). It is assumed in it that structural changes in agriculture call for transformations in rural areas. They are to be promoted, first of all, by measures improving living and working conditions in the village. A feedback between modernisation of agriculture and multi-functional development of the village implies that modernisation releases a part of work force from agriculture, which can be employed in non-agricultural professions (non-agricultural production, services, rural tourism). For the structural policy towards agriculture and rural areas to be effective it has to be comprehensive and internally cohesive. It means that it has to reconcile production, economic, social and ecological goals. Thus, a cohesive structural policy of development of rural areas and agriculture should be oriented at improvement of the following structures:
(i) employment in rural areas, that is, increasing employment opportunities outside agriculture creating, simultaneously, jobs allowing to supplement incomes from agriculture (part-time farming);
(ii) education of rural population perceived in such way that possessed qualifications and skills should allow to work effectively in agriculture, find employment outside agriculture or combine these two types of employment;
(iii) technical and social infrastructure allowing rural areas to become a favourable location for different types of economic activity and an attractive place of dwelling;
(iv) agrarian structure including farm-size structure, structure and quality of agricultural production;
(v) organisational-economic ties between agricultural producers and the market of agricultural-food products;
(vi) protection of landscape, natural resources and natural environment in rural areas along with their cultural heritage;
(vii) administrative and institutional structures promoting development of rural areas and agriculture.
Such cohesive structural policy of rural areas and agriculture development treats agriculture as one of possible economic activities in rural areas and a source of livelihood for a part of rural population. Economic instruments of the State assistance aimed at restructuring of farms and raising effectiveness of production will be addressed to those, who link their present work or future, to a large extent, with agricultural activity (developmental farms). Meanwhile, other instruments supporting development or launching of non-agricultural activity in the place of living, and also helping to find employment in new jobs created in rural areas and small towns will be addressed to the remaining group of village inhabitants finding employment in agriculture, who treat their farms (subsistence farms) as a source of products for their own consumption or as additional economic activity (and, thus, also an additional source of income). The State assistance will be oriented at professional retraining of persons seeking jobs outside agriculture. The State will be encouraging them, for instance, to benefit from early structural retirement and will provide funds allowing to give up agricultural activity.

From Agriculture to Rural Development in the Curriculum and in the Rural Economy

Andrew Errington

Seale-Hayne Faculty, University of Plymouth, Newton Abbot, Devon, TQ12 6NQ, UK.

Introduction

I want to start by defining the terms I am going to use in this paper. There are four key words in my title. By <u>agriculture</u> I am referring to the many different ways in which human beings manipulate biological systems in order to meet their own needs. It happens to be the case that in Europe at present most agriculture is geared to the production of food and fibre products and that the farm family business is the most prevalent unit through which this is done. While some of these businesses are very large and employ hired labour, others are small, relying entirely on family labour. Their key characteristic is that day-to-day managerial control is in the same hands as the ownership of the main farming assets. In theory at least, this gives them a significant degree of independence in their decision-making and the ability to respond very quickly to changes in their circumstances. Since many of these businesses are relatively small, the evolution of European agriculture ultimately takes place as a result of the decisions of millions of individual decision-takers.

However, they do not operate in a vacuum and their decisions are strongly influenced by the actions of a much smaller number of people - the managers of multinational corporations, some key politicians and civil servants, the leaders of some influential non-governmental organisations (NGOs) and so on. Their businesses, Government Departments and organisations form part of the wider agricultural system and they seek to influence the farmer's decisions in a variety of different ways - to buy particular machines, to sell their product to particular companies, to change established farming practices, or to give up farming altogether.

Agricultural colleges, universities and university departments have always sought to cater for the career development needs of the whole range of individuals involved in the wider agricultural system, not simply those who operate farm businesses. Fifteen years ago, when I carried out a survey of our own agriculture graduates from Reading University I found that roughly one-third went on to work on their home farm and one third to work elsewhere in the wider agricultural system. The remaining third used their degree as a stepping stone to a career outside farming altogether.

By the term <u>curriculum</u> I am referring to the content or subject matter of Higher Education. Working here at the Seale-Hayne Faculty of the University of Plymouth I am particularly conscious of all the developments currently taking place to devise new approaches to learning, changing <u>how</u> our students learn. But I shall be concentrating on content rather than method of delivery, while recognising that at least some of my points may have implications for method as well as content.

My own definition of <u>rural development</u> is quite short. In a paper delivered to an EAAE seminar in the Netherlands (Errington, 1999) I defined it as "premeditated changes in human activity which seek to use resources within the rural arena to increase human well-being". I went on to describe that arena, largely in spatial terms. By rural development I am therefore talking about deliberate change which seeks to make better use of all types of rural resource – land, water, buildings, people, landscapes etc - in order to increase the well-being of humanity. Of course this is a decidedly homocentric view (with which some in this audience will not agree) but it does emphasise the importance of more than just Gross Domestic Product as it is conventionally measured. For example, in my view those changes in human activity which safeguard or enhance environmental quality and in that way increase human well-being are just as much part of Rural Development as those which provide jobs or increase incomes.

The farmer who begins to add value to the farm's outputs by processing them before sale may therefore be involved in rural development, as is the farmer who decides to reduce the spring stocking rate on upland pastures, perhaps encouraged by an agri-environment scheme. But there are a lot of other people in society whose actions can contribute to rural development, many of whom have little if anything to do with farming. (This fact seems to be largely ignored in the Rural Development Regulation as it is currently constructed.) The graphic design firm which starts up in a small Berkshire village, the freelance copyeditor working from home in rural Oxfordshire, the Classic Triumph motor cycle spares dealer who has moved his mail-order business from a large city in South East England to a farmhouse in Devon, perhaps even the village hairdresser who employs the wife of local farmer are all potentially involved in rural development. And it is not just the actions of business people who may contribute to rural development. A whole range of community groups working to improve the facilities available in their village or market town may also be involved in "premeditated changes in human activity which seek to use resources within the rural arena to increase human well-being".

By now it will come as no surprise to you that I use the term rural economy to include all types of economic activity taking place within rural areas. In my view it goes much wider than farming or even land- or water-based activities. It encompasses the whole range of increasingly diverse economic activity than can, and increasingly does, take place in rural areas. As citizens concerned for the future of our farming communities it is important that we accept this broad view for it is very unlikely that farming families will be able to maintain (let alone increase) their standard of living if they rely on agriculture alone. As Higher Education establishments faced by a substantial decline in the numbers seeking to enroll in our traditional agriculture courses, it is also important that we do not think too narrowly for these other areas of rural development may provide us with some of our future markets.

Thinking about the market

In a previous paper for what was then the European Journal of Agricultural Education and Extension (Errington and Harrison-Mayfield 1995) we examined the impact of structural economic change in rural England on the future of our county agricultural colleges. At that time we pointed to a major reduction in the number of young people seeking to take courses in agriculture - between 1984 and 1990, the number of regular whole-time workers under the age of 20 engaged on agricultural and horticultural holdings in England and Wales had almost halved. We also described changes in the funding arrangements for the agricultural colleges which, from 1st April 1993, were removed from the control - and financial responsibility - of local government. On that date they become independent corporations with the status of charitable trusts, effectively competing in a national market for a declining number of students, but at the same time were given much more freedom to decide and develop their future role.

We outlined the four strategic options which they, like any other organisation facing a fundamental change in its market, could consider (see Figure 1). In the first place they could concentrate on selling their existing products to a different market, perhaps looking to sell a higher proportion of their product overseas (Option C). Second, they could switch their resources to the production of entirely new products for their existing markets where they are already well-known (Option B); or they could develop new products to sell into new markets (Option D).

Figure 1 The Strategic Options

Option A	Option B
Existing Products/ Existing Markets	New Products/ Existing Markets
Option C	**Option D**
Existing Products/ New Markets	New markets/ New Products

Some, we argued, would seek to survive by increasing their share of the declining traditional market, perhaps using more aggressive marketing tactics to win over customers from their competitors (Option A). Since change is both costly and uncomfortable, we felt that too many were likely to adopt this latter strategy and while some would survive, others would not.

By the time our article was published in 1995, different agricultural colleges in England were already exploring each of the strategic options in Figure 1. Many had diversified into the provision of new courses, though mostly for the land-based sector. As Harris (1993) showed, between 1984 and 1991 the proportion of all full-time students at agricultural colleges in England studying subjects classified as "Farming" fell from 54 per cent to 30 per cent. Other colleges sought to develop links overseas, catering for the vocational education/training needs of agriculture in the countries of the South or of central and eastern Europe. Some sought to win an increased share of the declining market for "traditional" courses in agriculture and horticulture. Yet others had moved towards Option D and had begun to explore the much wider rural market.

Further and higher education establishments certainly have an important role to play in facilitating the rural restructuring process in which the conventional production of bulk agricultural commodities has come to represent an ever smaller proportion of economic activity in rural areas. We can provide support for the positive aspects of change, facilitating economic expansion by providing the new rural business ventures with the skills and services they require. We can counteract the negative aspects of change; helping people cope with redundancy from sectors in decline by rebuilding their confidence and providing retraining to help the transition to new occupations.

My basic contention is that whether Higher Education itself is funded from the public or the private purse it must respond to the changing needs of the market if it is to survive and prosper. The economic and social restructuring of rural areas means that all establishments involved in Agricultural Higher Education will face a constant pressure for change in the curriculum, both in the content of the courses they provide and the types of course they offer.

A changing market

If we examine the main changes that are taking place in farming itself, we can identify the sorts of need for which Higher Education must cater in the future. Inevitably, because my own research, teaching and outreach activities relate primarily to the UK, the picture which I am going to present tends to reflect the particular circumstances of this country. Nevertheless, I hope you will find it useful to compare this picture with your own territory and see if you can recognise any of the five types of farmer that I see developing in the 21st century.

I have argued elsewhere (Errington 2000) that when we look at changes in agriculture we often overstate the importance of policy and technology and underestimate the effect of the deep-seated social and economic changes. Over the past 30 or 40 years, social and economic change has transformed the context in which the average English farmer operates. And it is the continuing evolution of this context far more than any change in technology or public policy that will continue to change the role of the English farmer into the 21st Century.

Over recent decades there have been four particularly significant changes influencing the shape of English farming. Each gives rise to a different potential role for the farmer. These changes are:

- "counter-urbanisation", the reversal of the centuries-old pattern of rural-to-urban migration, giving rise to the role of *farmer-as-incomer*;

- the beginnings of a "second wave" of counter-urbanisation in which jobs and economic activity are flowing back into rural areas creating a "new rural economy", giving rise to the role of *farmer-as-rural-entrepreneur*;

- a fundamental change in the public perception of the farmer's role in relation to the natural (and semi-natural) environment giving rise to the role of *farmer-as-countryside-manager*;

- changes in the markets for agricultural commodities giving rise to the roles of *farmer-as-commodity-producer* and *farmer-as-outworker*.

Counter-urbanisation: the farmer as incomer

In the early 1970s evidence began to emerge from several countries of the beginnings of a reversal of historic population trends in which rural areas tended to experience continuing population loss while urban areas expanded. In Britain, for example, the Population Census showed that while the most rural areas had experienced a population growth rate 5.5% below the national average.

As Selman reports (Errington et al 2000: Appendix 1):

"the 'remoter, mainly rural' districts....were the only district type recognised by the Office of Population Censuses and Surveys (OPCS) to have increased their rate of growth between the 1960s and 1970s, moving against the national trend; indeed, between 1971 and 1981 this district type recorded faster growth than all the others except districts containing New Towns. More detailed analysis confirmed that this strong performance by rural areas was not confined to more accessible places or a particular part of Britain but was extremely widespread, with some of the largest upward movements in change rate occurring in the most rural and remote areas".

The population turnaround in this country has been associated with a significant increase in personal mobility, arising from increased car ownership and a vigorous public policy of road building and road improvement. In 1965 36% of households owned one car and 5% owned two. By 1990, the proportion owning one car had risen to 44% while 19% owned two (Department of Transport 1993). The twenty years between 1965 and 1985 saw a 61% growth in total person-km travel, the largest proportion of this being attributable to longer journeys. Root et al (1995) estimated that the average journey to work doubled between 1977 and 1995.

Rural areas (and the farms within them) have thus become increasingly attractive as places to live, particularly for commuters and retired people who can now experience the benefits of living in what they perceive to be a more attractive location without sacrificing the benefits of urban jobs and facilities.

The process of counter-urbanisation has had some very significant implications for rural areas. It has introduced an influential new property-owning class who are not directly dependent on the rural economy for their livelihood. In some parts of the country, it has led to a substantial increase in the demand for farms as residential properties, incidentally providing significant underpinning to land prices even in periods of deep farming recession.

There is considerable variation in the impact of counter-urbanisation on different parts of the country. It may be most pronounced in the commutable "peri-urban fringe" surrounding the main metropolitan centres and in the most popular retirement districts. Where it has occurred, it tends to affect property and power relationships within rural communities. By increasing the number of incomer farmers whose objectives relate principally to the residential attractions of the farm it may also have changed not only the farming population itself but the nature and purpose of farming activities in these areas.

The New Rural Economy: the farmer as rural entrepreneur

As we have seen, the process of counter-urbanisation is fuelled by the residential preferences of the English population and their increasing personal mobility. The latter part of the 20th Century also witnessed a number of fundamental changes in the economic field, which have begun to transform the rural economy. Changing production technologies, the growth of the service sector, the downsizing of many large firms which are now placing greater reliance on the "outsourcing" of intermediate inputs in preference to their "in-house" production have led to the emergence of a multitude of small "footloose" firms which can locate themselves anywhere in the country. For many of them, the attractions of a rural location (perceived lower ground rents and access to a pool of compliant, low-wage labour as well as the intrinsic attractions of a more attractive working environment) have begun to outweigh the disadvantages.

These trends have given rise to what has been termed the "second wave" of counter-urbanisation in which jobs as well as people have moved in to rural areas (Newby 1985, Fothergill, Kitson and Monk 1985, Errington 1987, Breheny et al 1996, NRLO 1998). Some of the new rural firms have brought their own management and staff with them, contributing to the population changes described above. In other cases, the firms have been formed by previous in-migrants who have chosen to stop commuting and to become self-employed, working from home or the local village. Yet others have been established by former employees made redundant by down-sizing firms and organisations, perhaps now supplying the same services that they had previously provided in-house to the very firms that are now their customers.

Once again, these patterns are not universal and different rural areas have been affected differently. The result is a much more diverse range of economic activity across rural England than we have seen for many generations. This is illustrated in figure 2 which shows the employment structure in six different types of district in terms of the broad categories based on our Standard Industrial Classification (SIC)[1]. It will be seen that there is relatively little difference in economic structure between the different types of district, except for employment in agriculture, fisheries and forestry (which rises to 5.2% among the Remote Rural districts) and employment in banking and financial services which tends to be lower in the Remote Rural and Coastal districts. In terms of their broad industrial structure, the more rural parts of England are very similar to the more urban areas. Of course these districts do contain small towns but even when a finer definition of rural areas is used the economic activity of rural areas is not substantially different from that of urban areas in these terms (Errington 1990).

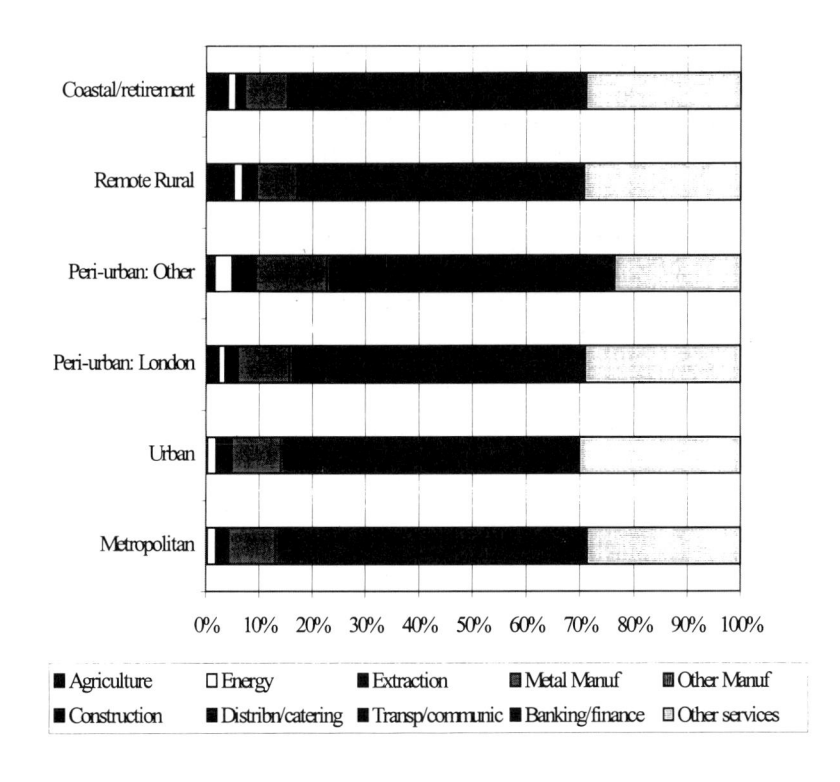

Figure 2 Distribution of Employment by SIC Division

---

[1] This analysis uses an adapted version of the typology of English local government areas developed by Hodge and Monk (1991) – see Errington (2000, forthcoming).

When the broad SIC "divisions" are disaggregated to the more detailed "classes", some interesting differences do begin to emerge. Table 1 reports only those SIC Classes within which 2% or more of the total workforce of at least one of the six types of district are working. It shows that:
- the "peri-urban: other" districts have a higher-than-average proportion of employment in traditional manufacturing sectors such as mechanical engineering, textiles and footwear
- both types of peri-urban district as well as the coastal/retirement districts provide a higher share of employment in electrical/electronic engineering
- the remote rural districts have a higher proportion of employment in food, drink and tobacco manufacturing
- both the remote rural districts and the coastal retirement districts have a higher proportion of employment in construction
- both the remote rural districts and the coastal retirement districts have a higher proportion of employment in hotels and catering
- the coastal retirement districts have a somewhat higher proportion of employment in retail distribution.
- all the peri-urban and remote rural districts tend to have a smaller proportion of jobs in banking, finance and business services but in the peri-urban districts around London, the proportion of jobs in business services is actually higher than it is in the districts classified as "urban".

As we look towards the 2001 Population Census it will be interesting to see if those sectors of the economy most characterised by footloose firms (e.g. electronics, business services) do increase their contribution to employment in the peri-urban fringe or indeed move out to some of the remoter rural districts.

What does this analysis of the broader rural economy suggest for the future of farm businesses and farming families? The "second wave" and the technological and economic changes that underlie it suggest new opportunities for farmers, perhaps converting and renting out farm buildings as attractive workspace for footloose firms. Where planning regulations permit, they may even begin to create a 21st Century version of the farm steading in which clusters of small businesses congregate in buildings around the farmhouse. The new rural economy may provide more diverse business opportunities for the farming family themselves as rural entrepreneurs. In some cases the new business ventures will be linked directly to farming; in others they will not. For example, some farming families will seek to capture more of the value that is currently added to their products elsewhere in the food chain, perhaps by group marketing, direct marketing, food processing or the production of branded products for niche markets. Others may simply run a completely new business on or from the farm.

Many farm family members will certainly look to the more diversified rural economy to provide them with full-time or part-time employment opportunities off the farm to supplement their farming income. Indeed this may be a much more attractive proposition than farm business diversification with its capital requirements and associated risks.

Table 1   Proportion of the workforce in selected SIC Classes

Standard Industrial Classification: Class	Metro-politan	Urban	Peri-urban London	Peri-urban Other	Remote Rural	Coastal Retirement
01 Agriculture and horticulture	0.3%	0.2%	2.4%[1]	1.6%	5.6%	3.8%
25 Chemical industry	1.1%	1.1%	1.4%	2.3%	1.3%	0.9%
32 Mechanical engineering	2.8%	3.1%	3.1%	4.9%	2.8%	2.1%
34 Electrical/electronic engineering	1.8%	1.7%	3.3%	2.9%	1.6%	3.1%
41 Food,drink/tobacco manufacturing	1.8%	2.1%	1.8%	2.6%	3.2%	0.9%
43 Textile industry	0.7%	0.8%	0.2%	2.1%	0.5%	0.2%
45 Footwear/clothing industries	1.3%	1.1%	0.5%	2.3%	1.0%	0.6%
47 Manufacture of paper/paper products, etc	2.3%	2.2%	2.3%	2.4%	1.8%	1.4%
50 Construction	6.6%	6.2%	7.9%	7.5%	8.1%	9.7%
61 Wholesale distribution (not scrap/waste)	3.6%	3.5%	4.4%	3.9%	3.7%	3.1%
64 Retail distribution	10.2%	11.5%	10.7%	10.3%	10.6%	12.4%
66 Hotels/catering	4.2%	4.2%	4.2%	4.2%	5.6%	8.0%
72 Other inland transport	2.5%	2.0%	1.8%	2.5%	2.4%	1.8%
79 Postal services/ telecommunications	2.5%	2.8%	1.9%	1.0%	1.5%	1.3%
81 Banking/finance	3.7%	2.9%	2.0%	1.5%	1.7%	1.8%
82 Insurance, not compulsory social security	1.5%	2.7%	1.6%	0.6%	0.6%	0.9%
83 Business services	9.2%	7.1%	8.1%	4.8%	5.0%	5.4%
91 Public administration, national defence, etc	6.6%	7.5%	6.9%	4.4%	7.1%	4.4%
93 Education	6.0%	6.4%	6.3%	6.0%	5.8%	5.8%
95 Medical/other health: veterinary services	6.0%	7.4%	5.1%	4.6%	5.6%	6.7%
96 Other services to the general public	3.9%	3.6%	3.3%	3.3%	3.8%	5.5%
97 Recreational/other cultural services	2.9%	2.1%	2.0%	1.6%	1.9%	2.6%

[1] shaded cells are those referred to in subsequent text

Changing public perceptions: the farmer as countryside manager

The last two decades of the 20th Century witnessed a fundamental change in public perceptions of the farm sector and its contribution to the well-being of society. Such perceptions are particularly important to an industry where so much of its income comes from the public purse. This makes the well-being of a significant proportion of farm businesses very susceptible to changes in the prevailing political consensus.

During the 1980s the public perception of the farmer as guardian of the natural and semi-natural environment of the countryside was seriously undermined. Public concern grew at the suggested harmful impact of farming practices on the flora, fauna and landscapes that were seen to embody the character of the English countryside. Typical of this period was an article in the *Observer* newspaper in 1985 headlined "Vandal Farmers" which asked "whether Britain's farmers, once the custodians of the countryside, have become its greatest enemy?" (and went on to argue that they had).

Reluctant to make greater use of regulatory powers and planning controls except in particular cases such as the point-source pollution of water courses and the removal of hedgerows, government sought to use the two other traditional techniques for influencing the actions of its citizens. More resources were devoted to the provision of information and advice to encourage the adoption of more environmentally friendly farming practices. And in 1985 the UK Government succeeded in persuading the then European Community's Council of Ministers to approve Regulation 797/85 "Improving Efficiency of Agricultural Structures" including Article 19 "Special National Schemes in Environmentally Sensitive Areas (ESAs)". For the first time in the EU, resources from the mainstream agricultural budget could be used to provide incentives to farmers to take part in environmental conservation activities.

By the time of the McSharry reforms in 1992, the identification of ESAs had become mandatory (under Regulation 2078/92) "in order to contribute towards the introduction or continued use of production practices compatible with the requirements of conserving the natural habitat and ensuring an adequate income for farmers". Even more significantly, European funding for the ESAs was to come from the Guarantees section of the European Agriculture Guarantees and Guidance Fund (EAGGF) which had previously been devoted to commodity price support. Agri-environment schemes were seen as a means of providing direct de-coupled support to farming incomes and had become an integral part of the Common Agricultural Policy (CAP). Indeed the planned allocation of funding to the various elements of England's Rural Development Plan (2000-2006) demonstrate that the agri-environment schemes are currently by far the most important component of the emerging "second pillar" of the CAP.

In essence, the agri-environment schemes recognise the role of the farmer as "countryside manager" producing a range of public goods - landscapes, flora and fauna - valued by Society (and for which Society is willing to pay). Though there continues to be debate over the levels and mechanisms of payment, the agri-environment schemes appear to have become an established part of the farming scene. But despite the plausible assertion that the demand for these environmental goods will rise as national income increases, there are still some uncertainties over the value that Society actually places on these goods, making it difficult to identify the point at which surpluses are created and payments should cease.

For some farmers, the agri-environment schemes are making them more aware of the environmental assets that they possess on their farms. Some are even beginning to recognise the opportunity to transform public into private goods. This might be done, for example, by exploiting the recreation potential of the landscapes and habitats created through the more environmentally friendly farming practices or by using these farming practices to produce locally-distinctive food products commanding a higher price.

Changing markets: the farmer as commodity producer and the farmer as outworker

My paper has so far identified three alternative roles for the English farmer of the 21st Century - that of "incomer", "rural entrepreneur" or "countryside manager". What then of the conventional farmer producing the traditional commodities of pigs, beef, sheep, milk, cereals or vegetables? There will still be a significant demand for these commodities in the 21st Century and a number of farmers will be well-placed to secure their position in an increasingly competitive market. They will tend to be the larger producers and their businesses will continue to grow, taking advantage of economies of scale. However, the nature of these scale-economies has now changed significantly, with important implications for many farmers.

For much of the 20th Century, the main economies of scale in farm production related to the fixed costs of machinery, labour and buildings. By spreading these over a larger number of acres the successful farmer was able to reduce unit costs and increase profits. As a result, most farm businesses grew through the purchase or renting of land adjacent to the home farm. The scale economies were place-specific. Acquisition of land even 30 miles away was problematic, often requiring a separate set of buildings and machinery, even a separate workforce.

As we enter the 21st Century, the most significant economies of scale relate to marketing and management. The buyers of the main agricultural commodities are looking for a homogeneous product of uniform quality delivered in bulk and on time for 52 weeks a year. Their own transactions costs are minimised by dealing with a small number of key suppliers. Where individual farmers have developed the skills and infrastructure to meet these exacting requirements they have a substantial asset. It is these established marketing networks and skills that now provide the key opportunities for scale economies as the commodity producer spreads marketing costs over a larger quantity of product. The commodity itself may be produced under a range of contract farming and sub-contracting arrangements in many different locations, with many of the sub-contractors effectively out-workers for the main farm business.

Some Implications for the Curriculum

I have described the more differentiated role of the farmer in the 21st Century as the conventional production of bulk agricultural commodities becomes less and less important in the context of broader rural development. I now want to draw out some implications for the curriculum. The fundamental premise of my argument is that the curriculum offered by Agricultural Higher Education needs to continue to evolve to meet the changing needs of the market. These needs will be reflected in both the design of new courses and the redesign of existing ones. Earlier in my paper I outlined the four strategic options that we face, including the provision of new products to new types of customer in a rapidly changing rural world. Already, many establishments, including my own, are exploring this option. For example, our Modular Masters Scheme targeted primarily at the post-experience professional updating market is offering programmes in "Rural Tourism", "Rural Property Management" and "Management and Marketing for Rural Businesses" as well as more traditional areas such as Agricultural Business Management.

But even if a Higher Education establishment chooses to concentrate only on its traditional markets – the farm businesses and the agro-ancillary industries that form part of the wider agricultural system - the changing roles of the farmer mean that some significant changes in the curriculum will be required. There is not space here to follow through all the implications of each of the roles that I have outlined in this paper but I hope that the picture I have drawn will stimulate discussion in your own particular subject areas. Perhaps three examples will suffice to start this process.

Many Higher Education establishments covering agriculture include farm business management in the curriculum. In some cases, and the Seale-Hayne Faculty is one of these, the curriculum covers the general principles and techniques of business management, using specific examples and case studies to tailor them to farming. Our students learn about the development of business plans but take into account both the objectives of the farm family and the resources of the farm. This "generic" approach to business management enables students to consider a variety of farm diversification possibilities such as the introduction of food processing or a farm tourism enterprise. However, if the farmer is to adopt the role of countryside manager, there will be an increased need to develop skills and understanding required to assess the environmental resources or "natural capital" of the farm. Moreover, since most of the direct payments for the environmental goods produced comes not from the market but from a variety of agri-environment schemes, the student also needs to have a good understanding of the origin and likely future of such schemes.

The current curriculum for some existing courses covers the construction and maintenance of farm buildings and their adaptation for use by a variety of crop and livestock enterprises. If it is to take account of the emerging role farmer as rural entrepreneur, this curriculum may need some adjustment. For example, it may need to cover the adaptation of farm buildings to non-agricultural uses as well as the legal and financial implications of renting them out as residential properties or workspace.

But perhaps most important for all our students destined to work on farms or in the agro-ancillary industries is the need to ensure that the curriculum allows them to develop an adequate understanding of socio-economic change in rural areas. As I have already argued, the changes currently affecting the farming community are not primarily driven by changes within farming itself and they cannot be understood by examination of agriculture alone. The curriculum studied by agricultural students must foster an understanding of rural development which extends beyond the farm gate and the agricultural sector. If it does not, there is a danger that the farming community will become increasingly isolated from the economic activities and social networks that in so many European countries already constitute the larger part of today's "rural world".

References

Breheny, M., Hart, D. and Errington, A.J., 1996. The Newbury Connection: An Economic Survey of West Berkshire, The University of Reading, Report to Newbury District Council 89pp.

Department of Transport, 1993. National Travel Survey 1989/91, London: HMSO.

Errington, A.J., 1987. Rural Employment Trends and Issues in Market Industrialised Countries World Employment Programme Research Working Paper WEP 10-6/WP90, Geneva: International Labour Office, ISBN 92 2 106336 4, 83pp.

Errington, A.J., 1990. Rural Employment in England: Some Data Sources and Their Use, Journal of Agricultural Economics, 41: 47-61.

Errington, A.J., 1999. Rural Development and the Rural Economist. In H. Hillebrand, R. Goetgeluk and H. Hetsen (eds) Plurality and Rurality: The Role of the Countryside in Urbanised Regions, The Hague, Agricultural Economics Research Institute (LEI), 1999 pp 115-133.

Errington, A.J., 2000. The Place of Farming in Rural Economy and Society, Journal of the Royal Agricultural Society of England, forthcoming.

Errington, A.J. and Harrison-Mayfield, L.E., 1995 Agricultural Education in England and the Dictates of the Market. European Journal of Agricultural Education and Extension, 1: 25-40.

Errington, A.J., Blackburn, S.P., Lobley, M with Winter, M. and Selman, P., 2000. Review of the Peak District Integrated Rural Development Project, 11 Years On. Report prepared for the Countryside Agency, Newton Abbot, Seale-Hayne Faculty, University of Plymouth.

Fothergill, S., Kitson, M. and Monk, S., 1985. Urban Industrial Decline: The Causes of the Urban-rural Contrast in Manufacturing Employment Change. London: HMSO.

Harris, A.G., 1993 Is Agricultural Education on Course? In P.Oakley (Ed) Farming Issues 7, Lloyds Bank, Bristol.

Hodge, I. and Monk, S., 1991. In Search of a Rural Economy, Cambridge University Department of Land Economy Monograph 20.

Newby, H., 1985. Rural Communities and New Technology, Langholm, Dumfriesshire: The Arkleton Trust.

NRLO, 1998. Rural Areas Put on the Map. Knowledge and Innovation Priorities. Aspirations for the 21st Century. NRLO Report 98/19E, The Hague, National Council for Agricultural Research.

Root, A., Boardman, B. and Fielding, W.J., 1995. The Costs of Rural Travel: Final Report, The University of Oxford Environmental Change Unit, Research Report 15.

From 'agricultural' to 'rural': the challenges for Higher Education

Keynote papers

Challenges for Higher Education in Rural Areas in the New Millennium in Poland

Andrzej Kowalski, Janusz Rowinski

Institute of Agricultural Economy, Polish Academy of Science, Warsaw, Poland

During the last two centuries economists argued that the economy functions on the basis of the three production factors: labour, capital and land. At the turn of centuries a theory of the knowledge-based economy has become more and more common. It means that the wealth of nations depends more and more on generation and proper use of the knowledge.

Ensuring equal opportunities at the starting point is one of the most important goals of the socio-economic policy in liquidation of unjustified social discrepancies between the urban and rural areas. The possibility for comprehensive development of people is valuable from an individual point of view; creation of optimal environment for meeting material needs, ensuring conditions allowing for self-realisation, full participation in the social life.

It has also its general social dimension. An individual with his experience, qualifications and needs is the main production factor of each society. The socio-economic changes in agriculture have an impact on the necessary structural adjustments of the labour force in agriculture. Efficiency of the structural adjustments in agriculture will depend to a great extent on the level of education of rural population. In this respect higher education will be of significant importance due to:
1. the processes of globalisation and integration
2. structural adjustments and restructuring of the Polish economy, including agriculture
3. civilisation backwardness of the Polish rural areas.

Even though the above-mentioned groups of factors exist in other European countries, Poland is a rather unique case. During the whole after-war period the advanced European countries were carrying out very costly adjustments in agricultural structures aiming, with various outcome, at transformation of traditional farm holdings into modern production units. From the other hand, in all centrally commanded countries – with exceptions of Poland and partly Yugoslavia – private farm holdings were transformed into the large scale state owned or "co-operative" units. In result after the Second World War significant structural changes were made in all European countries regardless of their socio-economic system apart from Poland. In case of Poland the clock for structural adjustments stopped. The lack of major structural changes in agriculture is one of the reasons for their disproportional development in regard to the whole national economy. Agriculture generates approximately 6% of GDP absorbing as much as 19% of all employed persons.

Labour efficiency in agriculture is 2.5-3.0 times lower than in the whole economy. This difference has its roots in the past. Unfortunately, the transformation period has not resulted in improved situation in this regard as the possibilities of employment in non-agricultural sectors of economy radically shrunk. A low level of education is one of the major reasons of lower labour efficiency in agriculture.

It obviously does not mean that farm holdings in Poland at the turn of millennium look the same way they did during the first years after the Second World War. Nevertheless, Polish agriculture must follow the same evolution path the Western countries went through during more than fifty years. Its pace though must be much faster despite considerably less favourable macroeconomic conditions. It becomes more and more obvious that the role of agriculture in the national economy is shrinking; its share in the generation of the national income is decreasing and technical progress, similarly to other sectors of economy, is of labour-saving nature. Smaller and smaller group of people can live from farming at the same standard of living as it is in the other sectors of economy. However, after the war, mainly in the 1950s and 1960s, quick development of non-agricultural sectors of economy was the main factor facilitating transformation of agriculture; part of the young generation quit the work in agriculture and easily found jobs elsewhere. In result, apart from production structures of the West-European agriculture also the character of rural areas was changed. At present, this factor has clearly disappeared and unemployment has become a real problem in the whole Europe. Unemployment in Poland will remain the key socio-economic problem in the coming decade due to very numerous young generation entering the production age as well as progressing restructuring of the Polish industry, of which rationalisation of employment is a part. It means that limited possibilities for restructuring of agriculture hinder improvements in the national economy's macro proportion due to the fact that maintaining high employment in the sectors of low labour efficiency makes acceleration of the development of the whole economy more difficult.

The issue of education in rural areas, not only higher education, should be considered in Poland in relation to the other European countries. It is known that in Poland only as many as 300-400 thousand families can live from farming in good living conditions. But there are about 2 million farm holdings, of which about half have no links with the market as they produce exclusively for own needs and the links of the remaining several thousand farm holdings are minimal. Problems of these several million group people will, as we believe, the most serious and difficult to solve socio-economic for at least the two coming decades. It is the more difficult to solve as, according to information provided in the further part of this paper, it is the least educated part of the Polish society, not prepared for appropriate functioning in the modern society.

The low level of education of rural population limits possibilities for finding jobs outside agriculture and hampers restructuring of farm holdings. The lack of significant changes in the employment structure influences low labour efficiency. Low labour productivity is accompanied by relatively low purchasing power of this group's incomes.

Appropriate preparation of the young generation entering its professional life is then a prerequisite for making necessary adjustments in rural areas in Poland. Hence, the issue of school education in rural areas is the primary problem that Poland has to solve as soon as possible. And higher education is not the most important in this regard. This is primary and most of all secondary education that determines rural society's readiness for functioning in the modern state.

Polish politicians are aware of this situation. The educational reform, embracing also rural education, is one of the four structural reforms having been implemented since 1 January 1999. Creation of the modern educational system that would fully meet the needs of modern society is the main goal of this reform. Obviously, transformation of traditional rural schools into the modern educational units will be the process lasting years, not only due to limited financial reasons, and its first results will be visible not earlier than in five years.

It is also obvious that shifting schools from their traditional "agricultural" character is crucial in the Polish rural education. It is worth mentioning that this evolution is taking place spontaneously. Rural schools at the secondary level, providing education in farm-related professions (so called agricultural technical schools), even though possessing good work conditions and highly educated teachers, suffer from lacking students who do not link their future with agriculture. Results of the questionnaire research indicated that not many children of farmers would like to run farms after their parents. It seems that it is a rather realistic evaluation of the situation. In this situation proper conditions, also intellectual, should be created so that the group of frustrated rural inhabitants is not growing. Even though they can look for new life opportunities in the cities it is known that newcomers from rural areas often have difficulties with finding their way in the cities. However, there is probably more favourable alternative – part of the rural young generation could stay in villages and work in non-farm professions. Obviously, it is possible only in "multifunctional" villages, although such villages may exist only if there are properly prepared people there. Hence the great role of education in rural areas.

In this regard an issue of the "rural " character of education arises. It is worth discussing this term as it can also be of negative nature. It should be noticed that agricultural education below the high level is usually evaluated negatively mainly due to the fact that it univocally determines the future of young people. Whereas in the modern society, in which as it seems it is necessary to change professions several times, an intellectual level that would facilitate fluent profession changes in both rural and urban areas is indispensable. Multifunctional villages differ from traditional ones as they provide jobs traditionally reserved for the cities. Therefore rural education should not limit people's mobility. It should prepare them for functioning in both urban and rural environment. Rural character of education should then be limited only to rural meaning in its positive sense.

Agricultural high schools face problems related mainly to matching the number of graduates with the real needs. It seems that in Poland, in which agriculture will remain one of the major sectors of the economy for the next several decades, higher education will play a great role. There is a room for about 300,000 to 400,000 farm holdings in Poland, which means that significant part of their owners should have higher or semi-higher education or at least obtain further education in the form of professional courses carried out by the university teachers. One could even risk the statement that the Polish agriculture, consisting of 300,000-400,000 economically strong farm holdings would need more people with higher education than it is the case now with the statistical number of farms of about 2 million. On the other hand, possibly fast adjustment of university teaching programs to quickly changing needs is indispensable. In this field, however, higher education in Poland reacts rather slow.

The level and types of education versus globalisation and integration

Integration of the Polish economy with the European Union and the world market is a natural and unavoidable process. Globalisation of the world market means among others gradual lifting of customs barriers and restrictions hindering the movement of labour and capital as well as production to the countries with lower costs. This process means also compulsion of radical improvement of Polish enterprises' competitiveness on the domestic and world markets.

Globalisation leading to the free movement over the limits of savings and goods, and in result investments and demand, results in the radical change of the criteria of efficiency. Globalisation not only lifts barriers limiting mobility of production factors, which is often neglected in its analysis, but also facilitates mobility of demand. An internal demand does not determine the demand for domestic products as its lack does not exclude production activity in the country, which is able to meet cheaply the external demand.

In this situation, comparative costs are becoming less and less important as the basis for specialisation in the foreign trade. It is the comparison of absolute costs that determines flows of capital and localisation of various production phases.

Totally free movement of capital, goods, services, labour force and information limits the possibility for isolation and special treatment of domestic production and demand as well as existing instruments supporting exports. In conditions of free flow of goods from abroad, competitiveness is sought both in exports and production for domestic market. Globalisation leading to the situation where everything can be produced everywhere stimulates changes in the character of labour division. The role of traditional specialisation in production of goods is diminishing in many areas, while specialisation in certain functions becomes more and more important.

Presently rigid boundaries of labour posts and hierarchical structures have started to disappear. Identification of the character of work by analysing the name of the position or title has become quite difficult. Tele-work, work at home, work and customer's sites, offices rented on hourly basis, staff and goods' leasing are the signs of disappearing of visible boundaries of organisations and admission of a new vision, in which professional life interlaces with a private one. Marketing rules have become binding not only in relations with customers, but also among the staff members.

The results of research made in various countries, also in Poland, aiming at collecting opinions of managers of large firms on necessary qualifications that employees should have in ten years time are very similar. The managers, almost unanimously, mentioned communication skills, and then ability to make right decisions, create professional relationships, sensibility to the firm's culture. More than a half of respondents listed practical skills as the last item.

According to the management staff, it is by far easier to train people in the field of technology or administration than to teach them social skills.

These opinions seem to confirm directions of expected changes on the labour markets. It is expected, for instance, that the share of workers on the American labour market will shrink from 20% to 10% in the coming decade, while the share of knowledge workers will increase from 60% to 70%.

The one who is able to take advantage of rarity and abundance relations becomes efficient and competitive. Chances for a long term success have only those countries and economic entities that possess resources of intellectual capital being able to use cheap resources spread on the global market.

The global market means not only severe competition aiming at maximum cost saving, but most of all competition in creation of the most attractive conditions for potential investors. Competition, of which the main goal is to attract capital, must bear various consequences. The set of most simple instruments includes tax reductions and granting firms with tax allowances. This, however, may have an impact on limiting state's possibilities for financing important social tasks, including educational ones. Besides, more and more often these tools become insufficient. The governments and societies for which social security is placed on the top of priority list become less competitive in relation to the societies, in which multi-generation family relations or social acceptance for limiting the scope of the social policy remain the main forms of social security. The liberalisation of capital flows results in the competition in lowering taxes, decreasing state expenses and resignation from social programs.

More and more often the pressure of financial markets and expectations of voters are in opposition. The logic of the global market indicates that capital markets become the key arbitrator in the economic policy as no one can afford to discourage domestic and foreign capital. The results of the above-mentioned changes are more painful for less developed countries, which have to liquidate the civilisation gap. They have to make up for the lost time in the field of infrastructure and education in order to become an attractive area for capital investments from one hand, and from the other hand, creation of attractive conditions makes it difficult to find funds for development programs and limits the range of instruments necessary for their implementation.

Effective allocation of production factors causes that the supply significantly exceeds possibilities of the efficient demand. Growing demand barrier results in not full utilisation of existing production capacity and this in turn leads to a huge unemployment amounting to hundreds millions of people. High rate of unemployment facilitates using the labour force in the system of "just in time worker". It means in practice replacement of full time and permanent employment by concluding short term contracts for doing a specific job or provide a specific service. This mechanism leads to a situation where employees risk loosing jobs in unfavourable market conditions. In addition, the costs of social insurance are incurred by employees themselves. In result only people with the highest qualifications have best chances for a permanent and long-term jobs.

In case of Poland consequences of liberalisation of capital markets are of various nature. Poland is rather poor in terms of resources important in the process of globalisation of economic activity. Production activities of firms with foreign capital in Poland are mainly capital- and labour intensive. Resources linked to advanced modern intellectual, production, research-development and infrastructural base are not sufficient for incorporating Polish firms to co-operation with the largest multinational corporations. Two traditional motives are the most important for foreign investors entering the Polish market, one of the market nature (size of the market) and the other of resource nature (low costs of production factors).

The process of integration will influence the situation on the labour market. In the structure of unemployment there will be, most likely, an increase in technological unemployment. In the European Union another restructuring of economy and the labour market structure are carried out. They are related to not only globalisation but most of all to the science and techniques. An increase in technological unemployment may be linked to the fact that creation of permanently competitive jobs will result in substitution of labour with capital both in the industry and in agriculture. More wide opening of the Polish economy for the EU and world markets will inevitably result in the market fluctuation unemployment.

Making the structure of the Polish economy similar to the one existing in the advanced countries will cause an increase in the structural unemployment.

Likely growth of trade deficit stemming from the integration process due to the reasons mentioned above does not have to negatively impact the dynamics of economic growth. If, for instance, net transfers of savings (such as BIZ) or resources from the EU structural funds are spent on activities related to accumulation or production investments, the growth rate may even increase.

The competitiveness of Polish products offered on the Common European Market requires urgent changes in the quality of Polish work. It is related to necessity of making the investments in broadly defined human capital more dynamic. Investments in this capital may at the same time increase the mobility of Polish labour force decreasing in this way the scale of unemployment, which in short run could result from the integration and membership in the EU. The European labour market should be considered from the Polish prospective not only as a place where one can find a job, but also as a place where Polish people can open service firms.

The perspectives of integration of the Polish agriculture with the EU imposes additional tasks on Polish farmers. Only professional management and economically strong farm holdings will have a chance to meet the competition of EU agriculture. The development of agriculture will have to be adjusted to the rules existing in the economies of partners and at the same time serious competitors. Meeting the requirements of partnership and competition demands from farmers thorough knowledge and understanding of the challenges as well as the character and forms of development factors in the modern world. Understanding of these mechanisms is a prerequisite for starting multidimensional adjustment processes. These challenges cause that schools have to teach pupils an ability to see own and local problems in the European dimensions.

An improvement in the level of education of rural inhabitants is one of conditions for absorption of financial and technical assistance from the European Union.

There will be no radical improvement in the competitiveness of the Polish economy without a priority for activities aiming at upgrading the state of education, as the competitiveness is determined in the long run by the following factors:
1. technical development
2. the economy is based to a greater extent on the development and use of the knowledge on how to develop creative and innovative skills
3. quality of work
4. restructuring of the economy aiming at increasing the share of services and high tech industries.

The necessity for improving the qualifications' level in Poland with the purpose to meet the international competition is of special importance in the light of some statistical data. The data clearly indicates relatively low level of education of the whole society, and its disastrously low level in rural areas (Table No. 1). There are significant differences in the access to education at its all levels in rural areas in Poland. According to 1998 data only 18% of children of age from 3 to 6 were covered by the pre-school activities (kinder gardens), while for Poland this ratio amounted to 28% (in the Western countries to more than 80%). Still more than half of rural inhabitants have only primary or lower education, which makes it a serious barrier in the process of modernisation of agriculture. This ratio looks even worse in relation to the comparison of the quantity of farm managers with higher agricultural education with the total number of users of private farm holdings.

A large distance is being maintained in the higher education in rural and urban areas. The percentage of urban population with higher education is five fold higher than of rural population.

Table No. 1
Education of population aged above 15 years according to the area of living (in %)

Type of education	Urban areas	Rural areas
Higher	9.8	1.9
Post-secondary	3.3	1.3
Secondary	30.8	14.1
Primary vocational	24.8	28.0
Primary	27.6	43.9
Not-finished primary	3.7	10.8

Source: calculated on the basis of "Rocznik Statystyczny" GUS

Educational conditions in the process of restructuring and modernisation of agriculture and rural areas

Education is a decisive factor in the quality of the "human capital", which is becoming an important factor of the economic and civilisation development. An access to education became one of the determinants of social and economic discrepancies. It concerns also agriculture. The knowledge used in practice is a base for the development of modern agriculture. A farmer's working place comprises of many natural, social, economic and technical elements. A farmer is both an entrepreneurial and executor, hence running a farm requires the knowledge and skills in many areas and at various levels of generalisation. In order to produce foodstuffs farmers have to know how to cultivate the soil, take advantage of weather conditions, what processes are taking place in grown crops and raised animals. The knowledge of soil requires acquaintance with chemistry, biology and crop mechanics as well as with elements of climate and meteorology sciences. Besides, it is necessary to know physiology of plants and animals in order to properly dose soil fertilisation and apply plant protection agents as well as to rationally feed animals or apply medicines and nutrients preventing diseases and facilitating growth. The knowledge of these processes is also crucial due to the quality of production and consumers' safety.

The technical knowledge of inputs, such as both tools, machinery (their construction and servicing) and chemicals (fertilisers, insecticides, pesticides, feeding preparation, etc.) is another area essential in the food production. Mechanisation and chemisation of agriculture requires more and more from farmers as without an efficient use of very advanced and complex production machinery and equipment it is difficult to imagine an efficient farmer, the more that the common knowledge is usually not reliable.

The third area of knowledge that farmers need includes economics and management, which allow farmers to run their farms profitably. The economic knowledge should help farmers to understand the market and enable them to calculate costs, etc.. The management skills facilitate better organisation of production, supply and sale, harmonisation of using labour and machinery, the more that work time cannot be regulated by the law or habits or delayed in time as the rhythm of nature must be followed.

The work of a farmer on his own farm holding nearly never brings a direct payment as it is always a type of a shorter or longer investment. Among the labour and machinery input and expected outcome there is an area of uncertainty related to both the nature and economic risk.

The economic knowledge of lower importance in closer or further past is and will play a key role in the economic and social standing of farm holdings and their development opportunities.

Running farms requires also the knowledge of social and political relationships, legal provisions, competences and procedures of authorities' activities and entities dealing with supply and procurement. Possessing those skills allows farmers not only to take part in the public life, but it is also a ground for setting development opportunities of farms. The political-administrative-social knowledge is crucial in some periods as it co-determines production decisions of farmers by influencing farmers' ideas on the future of farms and chances for implementation of their plans, aspirations and life goals.

The production of foodstuffs demands then a very broad knowledge. Agriculture is becoming a profession requiring more diversified education than it is the case in many urban professions.

Relatively low level of school education of rural areas' inhabitants is one of the most serious barriers hampering acceleration of structural changes in rural areas and agriculture. Only persons with high professional qualifications may count on finding jobs outside agriculture. Despite high unemployment rate in Poland during the period of transformation (10-16%), persons with higher education have not had difficulties with finding jobs. A low education level is one of those factors that have negative impact on entrepreneurship and innovation absorption. At the present stage of economic development of our country it is of particular importance due to the fast pace of changes in the economy. It is easier for better educated persons to follow and adjust themselves to those changes than for people with lower education. Secondly, the role of qualified staff is increasing in result of the present scientific-technical revolution as technological absorption of the region highly depends on the people. The absorption is then decisive in shrinking the technological gap among weaker and better developed regions. Shrinking of this technological gap depends on whether and how much the absorption is developed and whether it will grow or get weaker. Regions with low qualified people usually play the role of regions absorbing external innovations and are not their source. Hence they are in worse position the more that the low level of education limits also the scale of absorption. Decreasing educational backwardness of the population of problematic regions is one of the most important actions preventing their further socio-economic regression. The development of higher education is in this regard vital. It will depend on both the local authorities and on educational policy carried out by the state central authorities. The central authorities decide on opening of new high schools (both private and state) and on the size of resources allocated for the state schools. An increase in the technological absorption indicates a special role of technical high schools. They need more funds than for instance humanistic or economic schools. That is why there has been no private technical high schools opened in Poland yet.

The role of regional (voivodship) self-government has significantly increased in the development of education. Educational and research programs of high schools should be more closely correlated with regional programs implemented by the regional self-government. By co-financing of high schools (especially vocational schools) self-governments could influence (on the strength of concluded contracts) teaching programs in the particular region. In this way they could stimulate education in deficit areas and limit the number of students in those areas in which graduates would have difficulties in finding jobs. The educational policy of the central authorities and self-government cannot be limited to only higher education. The principle of autonomy is important due to not only tradition but as more than ten years of experience show to effectiveness of educational processes as well as creative competition among individual high schools.

Funds allocated for education, including high schools, are still low in Poland. The percentage of GDP spent in Poland on education is lower than in the EU countries, which enjoy significantly higher level of education and higher number of students per 1000 inhabitants (for instance in the middle of 1990 1% of GDP was spent in Spain and Germany, 1.4% in Ireland, 1.6% in Sweden and only 0.8% in Poland).

An improvement in qualifications is a process that requires fairly long period of time. Delaying in time actions aiming at acceleration of this process may become one of the main obstacles in structural changes of the national economy and improvement of competitiveness of the Polish economy as well as socio-economic development of the country.

Before 1989 an educational gap between the urban and rural areas as well as between agriculture and other sectors of the economy was enormous. In the period of real socialism education barriers in rural areas were not liquidated in spite of that fact that the scholarship system was much broader than in the period of transformation and privileges in selection to secondary schools were applied for children originating from rural areas (in the form of additional points). Delays in the economic and civilisation development of rural areas caused that young people that obtained qualifications migrated to the cities and took advantage of absorptive labour market in the industry.

Further differentiation of educational opportunities for rural children and young people took place in the period of economic transformation. During this time changes appeared especially in the functioning of pre-school child care units. These changes were of structural nature or resulted from different economic, social and demographic conditions of the social policy implementation.

Changes in the economic conditions relied mainly on limited financial capacity of the state, as the crisis in the public finance sector caused a fall in the budget expenditures for the social sector. The consequences were very severe as they were accompanied by a simultaneous increase in costs of running units rendering social services. The drop in funding education and child care from the public sources was compensated by an increase in family spending. According to GUS estimates expenses for education from the people's personal incomes account for about 10% of total expenses for this purpose[1]. The Ministry of National Education estimated that in 1998 17% of spending for the public and vocational schools was covered by the parents[2].

After the structural change of 1989 access of rural inhabitants to education worsened. The financial barrier and unfavourable structure of educational aspirations of rural population became more visible.

In the budget of households the number of children, their age, sources of income, place of living and education of the family head are the most important variables determining the level of expenses for education. Households of persons running own businesses add for education of their children from own resources about 3.5 times more, and employees about 2.5 times more than the households of farmers. Significant differences in the structure of expenses for education among other social groups exist as well. In the households of persons living from farming (more than 50%) expenses for school books and other items related to school education dominate[3]. Whereas expenses for non-school educational activities as well as the fees for social and private schools dominate in the budgets of households with non-agricultural sources of incomes. Differences existing in the level and structure of expenses for education in households running own businesses and employees as well as farmers and farm holding users are another evidence of limited access and not equal opportunities in the education of rural and urban children. Widening of differences in expenditures for education means deepening of differences in opportunities for better education of children and youth due to the wealth of the pupil's family. Limiting public funds for education and transferring of education costs to families resulted in making the opportunities of youth more limited and diversified.

Possibilities for further increase in household expenses for education as well as major expansion of private school education are limited by the low level of people's incomes.

The level of farmers' education is diversified regionally and depends also on the size of the farm.

Only small and very large farm holdings have ratios higher than the average for rural areas. In typical for Poland farms of 5-30 hectares only 1.5% of farmers have higher education. While on farms of above 50 ha the percentage of persons with higher education increases to more than 50%. The level of rural population education is strongly diversified regionally. The lowest level is represented by the population in North and East regions, while in central and Western regions rural inhabitants are relatively better educated. 58.6% of farm holdings' managers in Poland do not have higher education.

Productivity and economic results of farms are closely linked to the level of educations. In 1989 the final production per farm holding of farms run by persons with higher education was 1.7 times higher than of farms managed by persons with only primary education. In 1994 this figure increased to 2.3. In addition, on farms headed by persons with higher education there was higher accumulation rate than on the others. On these farms positive accumulation was registered even in the period of severe agricultural crises at the beginning of 1990s.

---

[1] Informacja o sytuacji społeczno-ekonomicznej kraju, rok 1994 s.70

[2] M. Gmytrasiewicz: Szkolnictwo zawodowe i wyzsze a rynek pracy. „Rynek Pracy" nr 5/1994

[3] Por. A. Baran: Wydatki publiczne i prywatne na edukacje. Wiadomosci statystyczne nr 6/1996

An improvement in education related to increased productivity of farming weakens traditional for small farms relationship between the size of the farm and intensity of production.

The level of education has direct influence on the pace and results of the application of technique and technological innovations. There is a strong relationship between education and how well farms are equipped with production inputs.

Effectiveness of education in Poland is very high, involved costs are giving return the period of 2-5 years.

Educational aspirations in rural areas, especially of farmers, are very low. Research done in 1984 showed that more than 15% of farmers are of the opinion that it is not worth getting educated. Farmers appreciate more the market value of education than its social status value. The motives for obtaining education include rather higher incomes than intellectual development or chances for social promotion or for learning about the world.

The lack of education deepens difficulties with understanding of changes and this limits chances for adjustment to new conditions. Soundings carried out among rural jobless people indicate that they consider the necessity of changing qualifications as social injustice.

Schools in rural areas are more poorly equipped than average schools in the cities. Only half of schools have gymnastic halls and football fields. A minor part has access to dining facilities, clubs or subject rooms.

Table No. 2
Percentage of population aged above 15 years with higher education in households running farm holdings, according to the farm size

	Total	1-2	2-5	5-10	10-20	20-50	50 and more
Percentage of persons with high education	2.01	3.33	2.07	1.33	1.19	1.94	9.37

Source: Powszechny Spis Rolny 1996

Demographic forecasts of Polish population provide that in the coming years the number of children and youth will decrease only slightly. More significant changes can be registered in particular age groups. In years 1995-2000 the population of children aged 7-14 years declined by almost 800,000. This downward trend will be continued until the year 2005. In the period of 2006-2010 the children's number should start increasing by 193,000.

The above-mentioned changes mean serious consequences for primary education, i.e. the very high number of pupils in primary schools in the previous decade (an increase by about 1 million in the age group of 7-14) will be followed by similarly large decline at the turn of century. At the same time the number of persons in older age groups will grow substantially (by more than 100,000 in case of 15-18 years old persons and by more than 343,500 in the group of 19-24 years of age), increasing the demand for room in high and post-secondary schools (in the 1990s the number of persons in this age group shrunk by almost 850,000), which than will again decrease by more than 1,200,000 by year 2010.

Systematically registered differences of increases in the number of children and youth in individual age groups, which are still the result of among others the Second World War, make it necessary to organise a flexible system of child care and education, as this is the only way of ensuring rational use of the units and proper education conditions.

Much more difficult work conditions and existing negligence demand particular qualifications of teachers and tutors working in rural areas. Unfortunately, the level of formal preparation of teachers in rural areas is much worse than in urban areas (Table No. 3).

Table No. 3
The level of teachers' education

Level of teachers' education	Total schools		Primary schools	
	Urban areas	Rural areas	Urban areas	Rural areas
Total	100	100	100	100
Higher	65.2	46.0	63.4	47.5
Post-secondary vocational	23.6	36.9	26.1	35.0
Secondary	11.1	17.0	10.4	17.4

Source: System ewidencji nauczycieli, MEN, 1998

The change of this situation, necessary for implementation of broadened teaching and educational programs, requires that new incentives (various forms of gratification and financial instruments) are applied, which would aim at attracting more graduates to start their professional work in rural schools.

The new 1990s generation of Poles have faced on its life path new development opportunities, but also very serious barriers at the starting point. It is especially visible at the moment of changing the position from a student to an employee. Large groups of young people have faced barriers in professional development and no job opportunities. The educational system proved to be not flexible and not able to adapt to the social needs in the market economy. The consequences of transformation have been observed also in other spheres of young people's life, such as leisure and spare time, cultural and family life as well as in models and social and political activity.

Education in Poland starts too late, i.e. at the age of 7 years, while in majority of OECD countries it is more and more common at 5 years (in obligatory system of schools, or in common pre-school system). The scope of information provided and required during the school year in Poland is broad (many experts consider it to be too wide), both in relation to children's perception and the content itself. It embraces a lot of news, even though the probability of using them in the decision taking process is rather low. From the other hand, many examples of important and useful information that schools do not provide and pupils cannot find easily can be presented. Equipping people with skills in searching methods and evaluation criteria are the primary goals of education. Teaching programs contain thousands pieces of information, but neglect almost completely the issue of methods for obtaining them. Polish teaching and training programs are not focused on how to get skilled and how to understand the surrounding world, but on obtaining a certain knowledge.

New goals the higher education faces makes it necessary to create the in-service training system more efficient. So far only 15% of population in Poland is getting additional education (in OECD countries 40%).

Results of comparative OECD research on functional illiteracy indicate that Poles are not very good in this respect as they cannot efficiently use available information necessary to function in the modern world.

Organisational and legal changes in the educational system in Poland

An improvement of the situation in education requires from the state a long term educational policy. This sphere cannot be regulated by only market forces, as the state intervention is needed as well. Education carries out wide and diversified tasks in the socio-economic and cultural development of the country and cannot be based only on signals the labour market sends. Education must create progress and a long-term development, which are not necessarily in compliance with present needs and interests of employers. The talents' selection policy should be implemented regardless of the place of living or property status.

Educational policy should specify the needs for certain professions as of today but also in the time of at least 20 years. This long term prospective in regard to teaching and education is essential due to very likely revolutionary changes that will take place in techniques and technology. The content and development directions of educational programs should be the result of diagnosis of expected changes in the economy and the labour market in Poland, the world market, opinions of employers on their activity priorities and requirements in regard to employees.

Multidirectional changes in the economy demand evolutionary transition from vocational education to education focused on obtaining professional qualifications combining general education and vocational education, as well as to module type of education that is more flexible, facilitate psycho-social activity and makes the knowledge supplementing more easy.

An increasing importance of the services sector causes that communication skills, including practical knowledge of foreign languages, and qualities necessary in efficient co-operation are becoming vital. This should be supported by linking vocational education, i.e. its content and structure, with the regional and local labour market. Self-governments of various levels having impact on the education at all levels should be responsible for education and training in their areas of responsibilities as well for initiating and monitoring of contracts concluded among employees, educational units and the labour offices.

As the experience of advanced countries show, the changes in education profile in rural areas will be facilitated to a greater extend by closer relationship of schools and local businesses. Production and trade enterprises and financial institutions operating in rural areas are much better oriented in the local conditions and real educational needs. Firms operating in large cities usually do not have such possibilities. Supporting talented youth, changing qualifications to own needs should facilitate not only financial needs of schools, but also program and staff co-operation.

So far employers in Poland have very rarely initiated professional improvement of their staff members aiming at better development of the company. Employers, however, face barriers in this regard in the form of organisational and financial conditions. There are various methods of financing training activities and courses all over the world. Training programs are financed from companies' own funds or from special funds created for this purpose and partly financed by the trade unions or employees themselves, public funds or special unemployment funds. Besides funds from various foundations and associations are used for this purpose as well. The system of tax allowances for employers organising or co-financing professional education is also well developed.

Acceleration of work on developing qualification standards for wide profile profession groups is another important task to complete. The profession qualification standards are necessary in the process of preparation of vocational teaching programs, as they will reflect requirements of employers. Without such a system accreditation and certification of both vocational teaching programs as well as units providing education and training would not be possible. The system of qualification standards should enable comparison of qualification skills of graduates of various vocational education forms, modernisation of teaching programs and vocational training. The qualification standards will improve reliability of certificates and documents confirming obtaining certain vocational skills. The establishment of a unified state system of vocational education will allow for coherence and flexibility of vocational education and will get the education in Poland closer to the EU standards. The implementation of qualification standards to the educational practice will demand a relevant system of teachers' education and monitoring of the teaching process.

The need for entrepreneurship development in rural areas obliges schools to undertake new tasks. Their implementation will be facilitated by the fact of taking over schools by local self-governments, but is not possible without creation of new type of teachers by high schools and without defining functions the educational units should play in local communities. They should play the role of institutions integrating the local communities, not only pupils. The teaching personnel should act not only as intellectual leaders, but also to a great extend as economic leaders. The establishment of intellectual infrastructure of the economic development of local communities is vital not only due to creation of the work patterns and pro-development and –innovative attitudes, but also to overcoming social disintegration.

Factors, which have recently become very visible, cannot be omitted while considering educational opportunities in rural areas. It concerns most of all more and more strong local communities' relationships, accenting own specifics, cultural or religious separateness. More active communities create in this respect more favourable conditions for better development of education in rural areas. Local communities may creatively and positively (but also negatively) influence the improvement of the authority of teachers and schools.

An access of youth to schools may be considered from the formal, social, spatial, cultural or economic point of view. In the studies carried out so far, a special attention has been paid to the formal aspect of accessibility to schools, related to the place of living or the region of origin, whereas various cultural aspects have been rather neglected in this regard.

The social environment is the primary factor differentiating conditions at the starting point of young people. Environment conditions, in which rural inhabitants live, are significantly different from conditions of living in urban areas. The style of work and living as well as food consumption habits, demographic and family structures are different. Agriculture is a very specific subject of the human activity as it is related to the biological processes and growth of live organisms. Whereas in the other sectors of economy these are material items, which are easier to handle, that people deal with. People can fully control the means and subjects of work, what is often not possible in agriculture. The work in agriculture can be characterised as very heavy, spatially unstable, periodic, dependant on nature conditions and risky from the nature and economic point of view. All these factors cause that there is a demand for a different labour force, different structure of employment, what obviously influence family relations, personal development, income level, consumption models, health, etc.

A considerable share of children's and young people's work is a characteristic feature of private family farm holdings. Comprehensive studies of pupils' time budget carried out in 1960s and 1970s by scientists from various scientific centres indicated that the work time of rural children amounted on the average to 15.5 hours a week and was increasing with the growth of age. The day of farmers' children often started at 5 a.m. with work before the school and ended at about 9 p.m.[4]. Even though the work of children is more and more often

---

[4] See among others: Z. Kwiecinski: Funkcjonowanie szkoly w srodowisku wiejskim. Warszawa 1972, J. Binczycka: Dziecki wsi i dziecko miasta. Wies Wspolczesna no. 9/1980, B. Tryfan: Praca dzieci w gospodarstwie rodzicow. Wies Wspolczesna no. 2/1996.

not treated as free labour force, the studies of time budgets of pupils of primary and secondary schools show that during the summer time the work on farms at the turn of 1970s and 1980s took from 3 to 5 hours a day. Shortening of children's and youth work time on farms reported in recent years by many authors results from both mechanisation of work, especially field work, as well as from following by rural parents the attitude to children of urban parents, characterised by more appreciation and investments in the development and education of their children. Hard physical work causes many losses and irregularities in the psychological and physical development of young organisms. Psychological and physical tiredness reduces general activity of young organisms, determining a low level of the final education. An engagement of children in work on farms results, in the best case, in shortening of their leisure and spare time. However, it often causes negligence of school duties by missing lessons and in result lack of promotion and not finishing the school. Also from educational point of view it brings more negative than positive results. As educators underline the above-mentioned losses are not compensated by such positive features as diligence, persistence, respect for hard work.

Assisting in work on farms from one hand and under appreciation of educational and cultural needs from the other hand determine opportunities for modelling personality of rural and urban young population and in result the life starting point. Young people that have to go to a far distanced school and assist on farms, that have no book collection and no access to consultants in the form of well educated parents are in a completely different situation than young people having at a close distance not only a school, but also a library, club, theatre, cinema as well as less duties at home and understanding of the importance of education.

Systematically observed significant differences in increases of children and youth in individual age groups force an establishment of the flexible organisational system of child care and educational units at all levels. This is the only method allowing for rational use of those units' capacity ensuring at the same time the proper quality of education.

Analysis of settlement network indicate that about 40% of villages in Poland have no conditions for opening a school as the number of children at one level is lower than 5. A possibility for organisation of a full eight-level school exist in only 20% of villages.

Scarcity of resources for education that the state budget and local self-governments face does not allow for soon opening in rural areas of schools applying new pedagogic trends. However, favourable climate and institutional conditions could be created already today and thus enabling opening of schools in the future, that would at least partly liquidate the demographic barrier.

There are, however, positive sides of educational units in rural areas in comparison with urban areas. They include among others ecological and health values of these regions, which in general provide conditions that are more friendly for human existence. Maybe, following an experience of some advanced countries, schools in rural areas become an attractive place for education of children from industrial regions, for whom ecological values are of great importance (air, water, food, recreation).

Lower crime rate in rural areas may become another positive factor facilitating changing the unfavourable trend of school "migration". Cities are more and more unfavourable for people as they facilitate various illnesses and provide opportunities for getting involved in the negative aspects of local communities' lives. Rural areas are still less suffering from social and civilisation pathologies. Education in rural areas provides opportunities for not only broader care or multidirectional educational and re-educational activities. Educational trends appearing presently in many countries and related to new substance and philosophy of education better suit organisational conditions of schools in rural areas than of urban schools. The vast majority of schools in rural areas are of small size, with a low number of pupils. This ratio is often quoted as a feature of a modern educational unit. Modern education or modern educational units are those that are small, have low number of pupils per class. However, in order to meet the challenges of education and care functions the schools must be equipped with modern equipment and staffed with highly qualified personnel. This imposes particular obligation on high schools preparing potential teachers of primary and secondary schools in rural areas. Besides, schools in rural areas provide better opportunities for more individual work with pupils and more flexible educational process. Small schools, low number of pupils, good knowledge of pupils' families, possibility for more full and permanent observation of pupils give an opportunity for better selection of talented persons and enables individual development work with them. Small number of pupils per class creates also better opportunities for those of them who have problems at school and may not get promotion. This is because they enable easier identification of problems and conditions for individual assistance. Individualisation of educational process allows for selection and individual work with all pupils.

Flexibility in the substance and process of education, care and teaching is one of the features of the modern educational system. Rural communities with smaller groups of people, who know each other, smaller schools and classes provide better base for handling quick and frequent changes, new information and knowledge inflow, necessity for interpretation of facts and views. They also allow for more quick change of the profile or teaching rules, moral models, mutual assistance and co-operation. All these opportunities that rural schools provide be noticed. So far some important possibilities for creating schools of educational pluralism were not given proper attention. There are great opportunities for opening non-public schools and other type of educational units (social and private) in rural areas. Spatial conditions (people's density, relationships among local community members) are favourable for establishing independent educational units in rural areas. Schools in rural areas are more concrete, more defined, more personal and their activities are more related to an individual person that is locally recognised.

Vocational education in rural communities may be more closely linked and subordinated to agricultural production conditions. Vocational education more and more often means permanent training as it covers people that already work, want to improve or change their qualifications according to the labour market needs. Schools linked with the region may very quickly respond to the local market needs by adjusting their profile to changing situation.

Taking into consideration all above-mentioned factors, being a certain type of relation between educational trends and educational conditions in Polish rural areas, the changes may facilitate liquidation of differences between rural and urban areas (cities-villages, big villages-small villages).

Educational policy should aim at initiation of educational aspirations of the society and mechanisms that would allow for meeting these aspirations. This goal has not been reached in rural areas yet.

The development of educational program for rural areas and wise support of youth capable of meeting social and economic challenges of the modern world, that would come back to villages after obtaining a degree and undertake activities facilitating restructuring of local communities, are the urgent tasks of the social policy. Teaching programs should be correlated with the development plans for rural areas that are adjusted to specific local conditions.

Conclusion

Lack of education is and will remain for a long time the most important obstacle in the development of Polish agriculture and its restructuring. Challenges the Polish agriculture faces are related to changing the approach of rural communities from the passive to the more creative and pro-development one, what can be achieved only through education. The lack of educational aspirations demand long-term and integrated activities of the state and social organisations.

The world progress in biologic sciences and technology resulted in a considerable growth of agricultural production. Plant and animal unit efficiency has increased. Food shortages has been replaced in some countries by food surpluses.

Agriculture and food economy create more commonly an integrated system of human activity aimed at production of full-value and safe food. At the same time rural areas and agriculture change their functions from producers of farm products to the management of food economy and development of services in tourism, leisure, recreation and hand craft.

An improvement in the people's standard of living by environmental protection and production of food of high nutritional, dietetic and treatment value are the crucial goals of rural areas and agriculture in the XXIst century. Requirements related to farm production will to a much greater extent take into account protection and recreational use of rural areas.

In order to achieve these goals the system of education, and most of all the higher education, need to change so that their directions and specialities are adjusted to conditions and needs of the present world.

From Agriculture to Rural Development: Critical Choices for Agricultural Education

Charles J. Maguire

Rural Development Department, the World Bank, Washington D.C., USA.

> The opinions in this paper belong to the author who is a Senior Institutional Development Specialist in the ESSD VP of the World Bank

Abstract

Making the leap from a concentration on production agriculture to a focus on rural development presents traditional agricultural education systems with difficult choices. The paper suggests that the choice is not whether to adapt to change but what changes to make. Rural development is a complex process that demands sustainable production agriculture, natural resources management, institutions, infrastructure, health, education, markets, finance, policy, local government, and education in order to succeed. Agricultural education systems from universities to non-formal adult education have to decide how much change they need to make to meet the expectations of an expanded and diverse population of stakeholders and remain relevant. Various authors have, over the past fifteen to twenty years, stressed the importance of institutional reaction to the pressures of change but action has been limited. The paper suggests that increasing competition from other educational institutions and non-traditional sources makes a strong and urgent case for agricultural education systems to make changes in order to influence a wide range of stakeholders including those in academia, in farming and non-farming rural areas, policy makers, and the private sector. The alternative is to become less and less relevant.

What is needed is the vision to sense the future needs of the multiple stakeholders in Rural Development and the leadership and determination to bring about change to enable the institution to educate, train, research and serve for the benefit of the rural community. The paper suggests that the time is right for change initiatives and identifies a number of international organisations and bodies that could be helpful to those contemplating systems change.

This paper is about change. Change we are told is the biological imperative and failure to heed the command can be fatal. Agricultural education is hearing the change imperative and needs to make some critical choices.

We have seen our education establishments develop and disseminate a number of truly momentous scientific advances related to plant and animal breeding, soil and water management, and food preservation. Our education establishments pioneered the development of ingenious labour-saving machinery; and introduced a business management approach to farming. Our education establishments have led the way in providing leadership in agricultural and rural sector policy formulation and reform. The agricultural education system has been a major contributor to agricultural research, extension, production and institutional successes over the past hundred years. We have much to be proud of but, at the same time, we need to be vigilant for the arena in which our success was won has expanded and changed. Alert members of the agricultural education community have not been unaware of the creeping and sometimes rapid influence of change and have made institutional and organisational adjustments to accommodate and remain relevant. However, we have reached a point where our change response has to be faster and more decisive if we are to survive as an influential force in agriculture and in rural development. Darwin awakened the world to the phenomenon of the survival of the fittest and many agricultural education institutions now face such a challenge. This paper suggests that it is time to take a measured look at the nature of change that surrounds agricultural education and resolve to adapt in order to take back a leadership role in education and training for a much-expanded clientele.

Signs of change

In many parts of the world, especially in Western Europe, North America, Japan, and Australia, the number of family farms has decreased to a level where commentators wishing to make a dramatic point suggest that farmers should be placed on the list of endangered species. Farming communities have shrunk as migration to urban areas has taken place. The size of viable farms in Western Europe, North America, and Australia has grown significantly, and the complex package of knowledge and skill required to make a living from the land includes a combination of plant and animal production, management, finance, marketing, science, and information. Agriculture has become a victim of its own success. Never before has production been so uniformly high yet there are pockets of hunger in both developed and less developed regions. Commodity prices continue to remain weak making the task of survival more difficult for all but the most skilled, resource endowed, and organised farmers. Consumers of agricultural products have changed from being content with a constant supply at a reasonable price to being knowledgeably critical of quality, price, purity and safety (see Box 1).

Box 1: An uneasy public (The Economist, June 19, 1999)

Critics assert that genetic engineering introduces into food genes that are not present naturally, cannot be introduced through conventional breeding and may have unknown health effects that should be investigated before the food is sold to the public . . . But there is a broad scientific consensus that the present generation of GM foods is safe. Even so, this does little to reassure consumers. Food frights such as "mad cow" disease and revelations of cancer-causing dioxin in food have sorely undermined their confidence in scientific pronouncements and regulatory authorities alike.

Society at large is conscious of the continuing assault on natural resources as forests disappear, water becomes polluted, or scarce, floods and landslides occur with greater frequency and ferocity and agriculture is classified as a major polluter and exploiter of natural resources. The profile of agricultural education students reflects the influence of change. There are now large numbers of urban born students enrolled in agricultural programs. The gender balance is more equal in what was a traditionally male area of study. There are decreasing numbers of the best students from secondary school entering agricultural degree programs, and demand for university graduates of agricultural education degree programs is down. Public support for agricultural higher education institutions is weakening. The locus of leadership in agricultural research and extension services to farmers has changed. The private sector has replaced or is replacing the public sector as the leader in making direct contact with the farmer. Private companies do research, produce inputs, run the "factory farms" and make money. Many of those who work for the private sector are not from agricultural education systems but appear to have the ability to make modern agriculture function and keep a reasonable balance between supply and demand for food and fibre. Is the agricultural education community aware of, and reacting to, these changes? Are the agricultural education and training (AET) providers viewing change from the perspective of institutions looking out from the comfort of an apparently safe haven or from the perspective of the client looking in with new demands and expectations? In view of the complex forces and elements which impact on things agricultural and, increasingly, rural do providers of AET fully appreciate that there is a new multi-faceted client population with different needs and very different expectations? Do providers have the imagination and the capacity to service the needs of the new clients? This paper attempts to highlight trends, contrast problems with opportunities and challenge the agricultural education community to develop future scenarios, act quickly or face a future without influence.

The implications of change for agricultural education

There has been no shortage of questions about the effectiveness of the traditional way of providing higher agricultural education and a range of responses has emerged from institutions around the world. We have seen the shift from pure production degrees to a greater emphasis on management, conservation, and agribusiness. We have seen a greater awareness of the need to conserve soil and water and the realisation that the traditional agricultural university cannot meet future rural challenges alone (see Box 2).

Box 2: The future of the United States' Land Grant College system

What can we expect to happen to the Land Grant system in the 21st Century? Slowly agriculture is losing its uniqueness. It is leaving the backwater and entering the mainstream, where it will have to learn to navigate. Public support for institutions that serve a diminishing number of people will decrease, as will the number of Land Grant Colleges. The agricultural disciplines – agricultural economics, agricultural engineering, agricultural biology, agricultural chemistry, agricultural business, and agricultural statistics – will gradually be absorbed by their parent disciplines. Agricultural colleges and agricultural courses will lose much of their uniqueness. Their research will become more interdisciplinary and large scale, with the agricultural components hard to identify. Agricultural extension will respond increasingly to the felt needs of off-campus people, from whom non-farm matters rank high.

Paalberg, 1992

Hansen (1990) notes that public sector institutions are not subject to the kind of market forces that govern the life of a firm. This is particularly true of agricultural universities, most of which are public institutions. In the absence of conventional market pressures, what might serve to ensure that the university addresses important social needs innovatively and responsively? Or, put in a more crudely negative sense, how does the university avoid stagnating and becoming irrelevant?

McCalla (1998) takes a world-view and warns that the agricultural science system will have to change. Isolated agricultural universities, dominated by the faculty and scientists, simply will not survive. The complexity of the challenge requires access to disciplines far beyond traditional agriculture. The changing role of civil society, participation and decentralisation will radically alter the clientele of universities and change the demands on them; and the role of proprietary private sector research will almost certainly increasingly dwarf public sector investments. Falvey (1996) echoes the theme of change by noting that a global reduction in the number of agricultural education providers should be expected. Some rationalisation may be seen in courses failing to adapt to changing requirements of funding resources and students. In other cases, it may seem a logical decision from the point of view of university management to fragment small agricultural faculties into their component disciplines within the faculties of science, social science and economics. Willett (1998) highlights the low level of investment in agricultural education and training over a decade of World Bank lending to Less Developed Countries (LDCs) and indicates that new approaches are needed. Engel and Wout van den Bor (1995) suggest that agricultural institutions are in flux. The relatively stable, straightforward institutional development they have known during the years since World War II seems to have come to an end. They no longer automatically form part of the mainstream of technological developments in agriculture. Besides, technical solutions are no longer sufficient. The further development of rural areas, including a responsible management and use of natural resources, requires social, economic, and organisational solutions as well. Moreover, shrinking government budgets and privatisation policies affect institutions' resource bases and their accountability vis-à-vis their clients, sponsors and society at large. Ruffio and Barloy (1995) raise the question that must be faced: how can one justify the existence of teaching and institutions specialised in agricultural sciences? The question has validity in the face of frontiers between disciplines disappearing, as demonstrated by the appearance of the bio-sciences – biochemistry, biotechnology, bio-mathematics, and bio-economics. They note that the replacement, or suggested replacement, of the traditional agricultural science degree by biology or bio-engineering diplomas is indicative of current thinking in some universities. Wallace (1997) notes the uncertain job market for graduates of higher and middle-level institutions which "will mean moving out of traditional production-oriented' agriculture into more managerial, entrepreneurial or non-agricultural occupations. Van Crowder and Anderson (1997) indicate that to maintain relevance, agricultural education institutions are now recognising that they must play an active and locally relevant developmental role as well as an educational one. Warren (1998) notes that life in tertiary agricultural education is tough, and will undoubtedly get tougher and that others cannot be counted on to help. He suggests that those who do not adapt will fall by the wayside. Paalberg (1992) observes that the experiment stations in the United States, which were responsible for agricultural research, began as almost the sole source of agricultural knowledge. They now share that role with biology departments of the Land Grants and other universities, agribusiness firms, the National Institutes of Health, the National Academy of Sciences, the International Research Network, independent research institutions and numerous agencies of the federal government – not just the U.S. Department of Agriculture. The discovery of new agricultural knowledge has, he suggests, become interdisciplinary, long-term, and expensive.

New Horizons for Agricultural Education

Virtually all commentators on the future of agricultural education stress the need for broadening purely production- focused programs in order to encompass rural development challenges that reside outside agriculture or are closely linked to it. When we envision rural development what do we see? The World Bank's strategy paper for rural development titled From Vision to Action (1997) noted that Rural Development encompassed all activities outside urban areas related to development. The strategy paper indicated that in future the Bank would be *taking a broad rural focus, as opposed to a narrow agricultural focus.* "The rural sector strategy focuses on the entire rural productive system. Water resource allocation and comprehensive watershed management incorporate irrigation and drainage. The management of natural resources in sustainable production systems treats agriculture, forestry, and livestock as part of a larger system. Human capital development, infrastructure, and social development are integrated into rural development strategies and programs."

Cleaver (1999) advised that to successfully support Rural Development the World Bank should in its lending expand the substantial content of non-agricultural rural development by giving equal weight in rural development strategy to rural education, health, infrastructure, non-agricultural economic activities, administration, community development. The Bank should also elaborate how governments, donors, the Bank will support investment in these activities. Some suggested interventions included:

- Rehabilitating integrated rural development programs
- Developing useable models of rural content for national education, health, and infrastructure programs.
- Developing national community development programs with rural focus
- Establishing Rural Social Funds
- Elaborating sub-sector investment and policy models of rural infrastructure, rural health, rural education, rural administration, and rural community development.

Wallace (1997) suggests that, in addition to specific skills, rural people are also seeking education for life which includes leadership development, confidence building and problem solving.

Hansen (1990) broadened the focus of agricultural education by noting that universities neglected the important policy and institutional variables that set the course and defined the parameters of rural change. Universities had been slow to explore how different rural strategies might enhance the generation and distribution of employment and income; strengthen the income-earning capacities of resource-poor households; improve the management of soil, water, agroforestry, and common property resources; and increase efficiencies, as well as equity in product and credit markets.

Britz (1998) also sees the importance of universities seeing agriculture in a broader sense and providing leadership for new stakeholders who represent the interaction between urban and rural cultures. He notes that farmer's political power diminishes and so does support for agricultural programs. However, urban people want resort areas, national parks, peri-urban and even urban agriculture. Greening of cities, preference for organic products or green political parties are also popular. Agricultural universities should analyse this movement with teaching and research activities, organising seminars and discussion groups.

Falvey (1996) suggests that the dual concern for the environment and the need for increased food production provide a context for future agricultural education. Existing courses mainly take a balanced scientific approach – to this there would appear to be a need to include a greater input from the humanities including an understanding of environmental ethics. The imperative to produce food, as far as we know to-day, will continue to rely on continued intervention in the natural environment. In accepting the responsibility to manage the environment with care, agricultural education may need to see itself as a field of natural resources management –managing the natural resource base (soil, water, mined fertilisers and so on) to produce food while understanding the interaction with that resource.

FAO (1995) suggests that the responsibility of the schools of agriculture should not be limited to turning out a professional elite with a strong scientific background but lacking the humanism essential to work in the social environment.

Rural Development takes place in a complex environment

In order to bring about significant change, reformers of agricultural education institutions or systems must appreciate the complexity of the environment in which a shift in focus from agriculture to rural development would take place. Within the rural development universe there are attractive and compelling activities which can be added to or incorporated into traditional agricultural education programs but many times these are selected not from a clear understanding of their place in that universe but for other reasons. Bawden (1998) offers a diagrammatic explanation of how the agricultural education and training (AET) system fits into the complex rural development system and how both of these systems are influenced by the wider environment in which they exist.

First, agricultural education is viewed not from a single perspective such as higher, secondary, vocational or adult but as a system. The Agricultural Education and Training (AET) system can be viewed as being composed of four inter-related sub-systems (Figure 1).

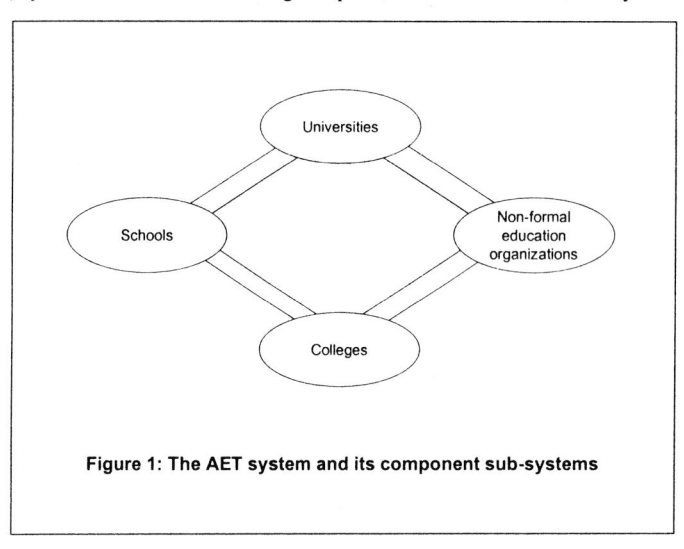

Figure 1: The AET system and its component sub-systems

The AET system is itself a sub-system of the higher order (agriculturally focussed) system (or bounded network) of rural development (Figure 2) which includes the clientele, organisations and institutions in both the private and public sectors, and both non-government (NGO) and community-based organisations (CBO).

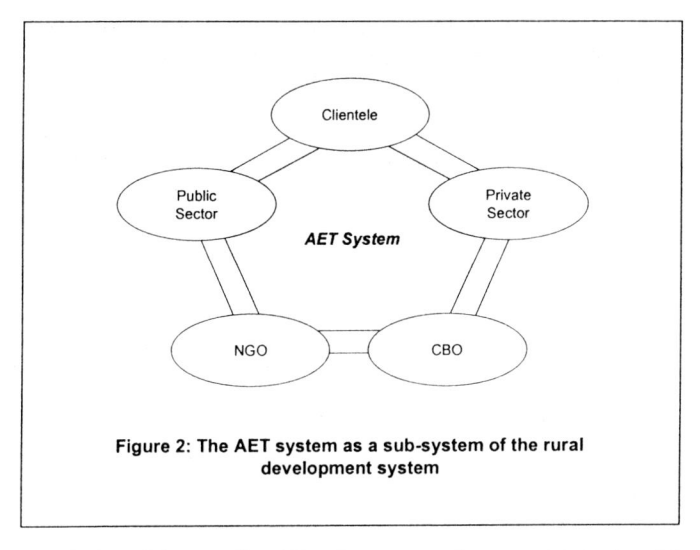

Figure 2: The AET system as a sub-system of the rural development system

The complexity of the term Rural Development can be appreciated when it is understood that each of the other sub-systems in the rural development system is also comprised of a complex set of interconnected elements. The clientele sub-system, for instance, can be seen to comprise seven inter connected sub-sub-systems (Figure 3).

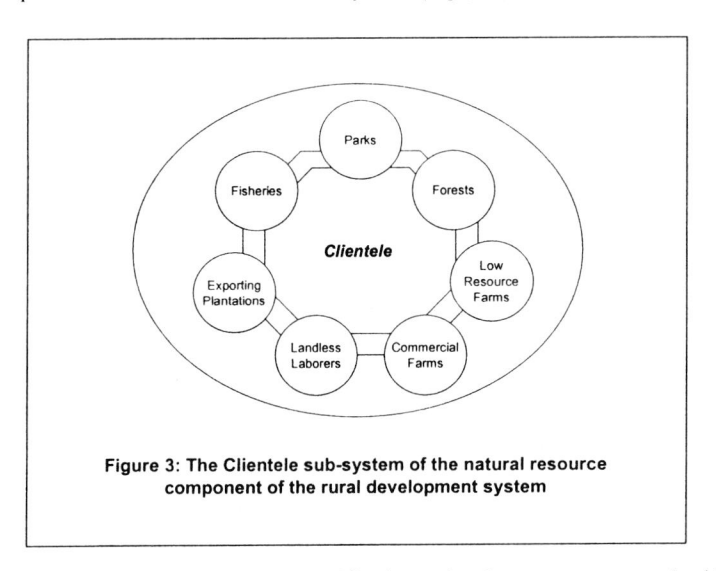

Figure 3: The Clientele sub-system of the natural resource component of the rural development system

The Public Sector sub-system meanwhile, has sub-sub-systems concerned with policy, infrastructure, research, development, and extension., while the Private Sector comprises financial institutions, input companies, marketing companies, manufacturing companies, the media, etc.

The whole Rural Development System meanwhile, with all these component sub-systems, itself operates within an environment of immense complexity which is characterised by a host of factors which can influence, and be influenced by, the rural development system (Figure 4).

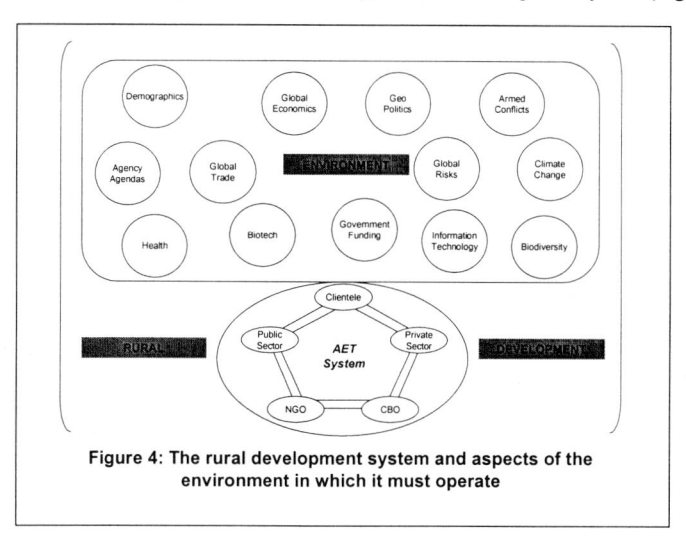

Figure 4: The rural development system and aspects of the environment in which it must operate

Taking a System-wide View of AET

We focus on higher education as our area of enquiry but increasingly we find it more difficult to separate out the various levels of education and training delivery in the pursuit of rural development. Higher agricultural education is, of course, a keystone in the system for it represents the locus of leadership- generation in policy-making, agricultural research, agricultural extension, production farming and teaching. We are correct in looking to higher agricultural education as the frontrunner in meeting the challenges of the future but we have to recognise that the neat divisions, which made our higher education focus legitimate, no longer hold. We know that one of the most critical job categories in agriculture to-day is that of the technician who can manage a farm, run a processing plant, repair agricultural machinery, market farm products, and provide goods and services. This essential category is not separate from the concerns of University level education. We see a decline in the teaching of agriculture in secondary schools where there are significant gains in the teaching of environmental principles. Surely we should be concerned that the place of agriculture in society and its economic, social and environmental role needs to be made clear to those who will shape the world in the early part of this new century. We cannot ignore the rising interest in rural education and in the search for links between rural and agricultural education (Sariego,1999). We have to be conscious of the increasing numbers of part-time students who seek knowledge and information related to agriculture and natural resources. We must view higher agricultural education as one element in an increasingly important system as we deal with rural development issues. It should be recognised that a shift in thinking to a systems approach will not be easy for all. Bawden (1998) suggests that a systemic perspective presents a complex challenge to those accustomed to thinking of agricultural education and training in terms of publicly funded universities focussed essentially on serving the needs of export oriented and commercial producers. Such institutions rely on publicly funded research and extension activities, and operate within relatively stable natural and social environments.

Donor Support and Agricultural Education Systems

For too long the focus of international donor agencies has been on elements of the AET system. The World Bank, in the twenty-six years between 1963 and 1989, supported agricultural education in sixty-seven of its 135 higher education projects. As a 1992 World Bank Review notes, agricultural colleges and universities were among the first education institutions to receive Bank assistance, and the Bank itself was among the earliest multilateral donors to support these institutions. Bank assistance was based upon the need to supply technicians to support the science-based agriculture, which was to play such an important part in increasing food security and promoting economic development. Governments looked to these higher education institutions to produce the technical personnel, managers, teachers, researchers, and extension workers required to staff agricultural agencies. It is now clear that the emphasis of these investments was on an element of the system, the higher education element, but not on the system itself. Willett (1998) noted in his review of agricultural education support by the World Bank and other donors in the decade 1987-1997, that past investment tended to emphasise bricks and mortar, hardware and faculty overseas training to build AET programs focused on state-led support services for production agriculture. The review identified the need to shift the paradigm for AET toward a much broader, multi-disciplinary systems approach. New generation AET projects need to develop human capacity, not just for production agriculture, but for environmentally and socially sustainable development throughout the rural sector, engaging more diverse, rural sector-related systems through a multiple field of partners and stakeholders.

Responses to the challenge of change

As previously observed, there have been institutional responses to the pressures and challenges of change and many of the responses have had positive impacts on aspects of higher agricultural education. A sampling of such responses serves to reflect the breadth of thinking about the future needs of the sector and how to meet these needs.

McCalla (1998) suggests a curriculum for the future at bachelor's level that would offer intensified production systems on two tracks. One would have a biological-physical emphasis, the other, a social-economic emphasis. All students would, in the first two years, be well grounded in fundamentals- mathematics, chemistry; physics, biology, history and philosophy and logic as well as modern information systems. Those on the biological/physical track would take a third year of molecular, cellular, physiological and systems biology; biochemistry; genetics; and anatomy and morphology. Those on the economic/social track would take a year drawn from anthropology, sociology, economics, political science, history, statistics and quantitative analysis. After the third year all students would begin the process of integration by a course of study focused around ecological and social systems. Using integrated production systems as a model- the program should focus on cross-cutting themes such as:

- Conservation Tillage
- Integrated Pest Management
- Integrated Nutrient Management
- Conservation, Ecology, Bio-diversity

- Biotechnology.

In addition, they would focus on farmer-farming system interactions and on farmer-village-social system interaction. All students would require an understanding of the importance of policy and institutions.

A fifth year would engage students in a field project as well as learning a foreign language.

The Master's curriculum would build on the field project of the fifth year of the bachelor's program allowing the students to focus more in depth on critical components of the cross-cutting themes. The program at doctorate level would be basically focused at the discipline level. The doctorate should remain a research degree.

Ruffio and Barloy. (1995) propose three principles which should guide decisions regarding student's future education. A strong basic scientific training, sufficiently broad-based to give future graduates a high capacity for abstraction, methods of reasoning and a greater facility in building theoretical concepts, explaining and stating terms of a problem and expressing technical questions in scientific terms. A more limited technical training, crucial indeed but not aiming at encyclopaedic knowledge. The development of personal qualities essential in position of responsibility: communications and managerial skills, ability to organise and adapt and to work hard. Wallace (1997) notes curriculum weaknesses in Africa which include a failure to embrace emerging global issues such as sustainability, the environment, gender, farming systems development and suggests a slowness in incorporating more transferable skills such as languages, computer studies, communication, marketing, and entrepreneurship.

In the U.S.A. which boasts a unique and much admired agricultural education system concern about the future led to the formation, in 1995, of the Kellogg Commission on the Future of State and Land Grant Universities. The Commission, funded by the Kellogg Foundation, was created to rethink the role of public higher education in the United States and had a time horizon of four years. The Commission was created in response to a public perception that the Land Grant system was out of touch and out of date and that expertise on campuses of the system was not able to focus on local problems. The Commission concluded in 1999 that the system needed to go beyond outreach and service to 'Engagement'. By engagement is meant a redesign of teaching, research, and extension and service functions to become even more sympathetically and productively involved with their communities, however community may be defined. Embedded in the engagement ideal is a commitment to sharing and reciprocity. The Commission envisaged partnerships, two-way streets defined by mutual respect among the partners for what each brings to the table. The Commission decided that the engaged institution must accomplish three things:

- It must be organised to respond to the needs of to-day's students and to-morrow's, not yesterday's
- It must enrich students' experiences by bringing research and engagement into the curriculum and offering practical opportunities for students to prepare for the world they will enter.
- It must put its critical resources (knowledge and expertise) to work on the problems the communities it serves face.

In parallel with the Commission and again with the support of Kellogg Foundation funding a national group undertook a multi-year (1996-1999) review of secondary level agricultural education with the task of Reinventing Agricultural Education for the year 2020. The motivation was similar to that which spurred the Commission, a fear that the secondary, Vocational Agriculture, program no longer reflected the needs of rural America and the challenges of globalisation and that unless it redefined itself it would be irrelevant in the education system. The outcome of the review suggests that "the work begun in this initiative

(Reinventing Agricultural Education) prepares us to deal effectively with changes in food production, environmental stewardship, human health, technology and the global economy."

In Australia, the University of Melbourne, undertook a rationalisation of the state system by uniting the university and a number of colleges in a single system.

A recent OECD Conference (1999) concluded that governments have tended to bring policy making for research, higher education and development/extension for agriculture and food into closer connection with general public policy for the development and provision of research, education and development services to their societies as a whole.

University College Dublin now has a Department of Agribusiness, Extension and Rural Development offering nine specialised degree programs. From the traditional Bachelor of Agricultural Science degree which focused on production agriculture there is now a choice of Agribusiness and Rural Development, Agricultural and Environmental Science, Animal and Crop Production, Animal Science, Commercial Horticulture, Engineering and Technology, Food Science, Forestry, and Landscape Horticulture. This major change from the traditional program was sparked. off by outside events. At postgraduate level, in addition to Masters and Ph.D. by research, the Department offers a Masters in Rural Development and a Higher Diploma in Rural Development. There is a continuous feed back on the effectiveness of degree programs from employers and graduates as a way of detecting change.

Choices for the future

Delbertin (1992) offers three options for the way agricultural education universities can face the future and speculates on the consequences of each.

Continue to Primarily Serve Commercial Farmers

Continue to devote most resources to output-increasing technical production research and rely on early-adopting farmers and their commodity groups for primary political support. This is a risky alternative given the declining number of commercial farmers, the decreasing comparative importance of the commercial farming sector, and the potential erosion of political support for output-enhancing agricultural research by politically more important groups such as urban consumers and environmentalists. The traditional target group cannot sustain agricultural research funding and downsizing must follow.

Build a Political Support Group Among Consumers

Continue emphasis within agricultural colleges (universities) on increasing output, but broaden the political and funding support base by embracing consumers who benefit from agricultural research in the form of lower food prices and a safer, higher quality food supply. Consumers, not farmers nor their commodity groups should be the political support base for the lion's share of agricultural research funding. Administrators have to seek out consumer advocates and provide them with opportunities to contribute to the setting of research agendas similar to that now given to traditional agricultural constituencies. Low income people would be the major beneficiaries of this research.

Emphasise the Problems of Non-Farm Rural Residents and Non-commercial Farmers

The third alternative is to redefine the mission of agricultural colleges (universities) to focus primarily on improving the well-being of all rural residents, farm or non-farm. Many would argue that this is now the case but a genuine focus will require substantial relocation of funds away from output-expanding research, and toward social and economic research dealing with the problems of non-farm rural people, poor as well as rich, and farmers that have not been a major source of political support for agricultural research. These include part-time farmers, organic farmers, and farmers with limited financial resources for adopting new technologies.

Consequences of these the second and third options could be that social scientists interested in improving the well-being of non-farm rural residents and farmers in these other categories could strongly influence the research agendas at agricultural colleges (universities). This option will meet with opposition from technical production scientists and early adopters. It is clear that as the number of commercial farmers declines, agricultural college administrators must build a broader political base of support to these other groups. Conflicts between what consumers want (as well as environmentalists, organic farmers and animal rights activists) and what is politically powerful and commodity-oriented commercial farmers want will be the primary administrative agenda item over the coming decade. Bringing both rural and urban consumers into the political support base could also have a high long run payoff in state and federal support for agricultural research. But first, consumers need to be convinced that they, not farmers, are the ultimate beneficiaries of most of the gains from agricultural research. Building coalitions will not be easy.

Partnerships

Advocates of change make the point that partnerships will be needed if progress is to be made. Partnerships within agricultural education systems; partnerships between agricultural education systems and larger education systems; partnerships with the private sector; partnerships with a range of other stakeholders including employers, non governmental organizations (NGOs), ministries representing other sectors, farmer's organisations, environmental groups, and consumers of farm products and services.

Topel (1998) notes the downward trend in financial support for agricultural research, extension and teaching in the United States and suggests that in order to achieve national and international recognition for quality research, teaching, and extension programs, funding from private individuals or companies is essential. Press and Washburn (2000) report that corporate giving to education is on the rise, growing, in the United States, from $850 million in 1985 to $4.2 billion less than a decade later. They also note that increasingly the money comes with strings attached.

Risks

If it is inevitable that Agricultural Universities and other higher education institutions dealing with agriculture and rural topics need to enter into partnerships with private businesses what are the risks that should be considered? (see Box 3) The biggest is that rural investments without the potential for profit may be ignored and neglected. Professors worry that universities will cease to serve as places where independent critical thought is nurtured. For example, raising questions about the safety of genetically altered crops may prove difficult if more and more agricultural colleges turn to corporations to finance their research (Press and Washburn 2000). The lesson from the American experience with partnerships is that the bigger partner may set agendas, influence the end product and, at the end of the day, be the big winner. The losers are the students and society at large who are deprived of a broader but non profit-making education and the curiosity that goes with research and developing an understanding of the world as it is. If we cannot avoid a one-sided partnership route to survive in an increasingly competitive education world maybe we should leave the non profit generating rural issues to others? Ruffio and Barloy (1995) suggest that in Central and Eastern Europe universities should become active partners in local and regional development, in partnership with local authorities and professionals from the business world, within structures yet be to be created. Press and Washburn (2000) ask the critical question:

"In an age when ideas are central to the economy, universities will inevitably play a role in fostering growth. But should we allow commercial forces to determine the university's educational mission and academic ideals?" W. van den Boer et. al. suggest that as long as institutions for higher agricultural learning are forced to attract more and more outside funding, they will not be able to afford the "luxury" of making clear and relevant choices about their societal mission.

The idea of partnerships within the greater education system may cause problems for established staff and institutions. While we like to think that partnerships are of the win-win variety it is clear that some wins are greater than others. It may be difficult for some agricultural education staff to give up identity by teaming with another academic program and similarly difficult for long established departments or faculties to face merger and loss of traditional "brand name". In the most critical circumstances the choice may be between merging and disappearing.

Box 3: A word of caution

Universities need to diversify their sources of information, enter into strategic alliances, be experimental, and mix boldness with prudence; they should also remember who must be in charge. The university must serve the future job market, not be a slave to it. There is a need to seek new relationships, not to give up one's ultimate responsibilities.

Dlamini, 1999

Why should we be concerned?

If the pressures for change are so great and if a future where the agricultural education that we know will be inevitably lost in a new and larger fabric of education for rural development why should we expend energy on dreaming up future scenarios? Perhaps we should resolve to shape the future rather than allow it to evolve in ways that may not reflect our judgement on how best to proceed. We have a proud tradition of contributing to the phenomenon of a very successful international agricultural industry, which has confounded the dire predictions of Malthus and others. Ours has been a major contribution in the areas of knowledge generation through research, education of countless thousands of personnel for the public and private sectors, agribusiness, production agriculture, and policy formulation. We have been the anchor for an agriculture when it was considered a way of life, and, later, when agriculture was recognised as a business. Now that the profile of agriculture has changed, to reflect demography, biotechnology, a greater emphasis on natural resources management, and a concern for the environment, we are in the best position to take a lead as the intellectual leaders and the developers of human capital for the next century? Not only is there an opportunity to claim a leadership role, there is also an obligation for as Conway (1997) reminds us there are some three-quarters of a billion people chronically undernourished and this figure is not likely to change over the next twenty-five years. If progress is to be made in reducing the numbers of undernourished we have to fill severe gaps on our knowledge. He suggests that scientists need to provide a better understanding of the genetic basis of yield, of such key processes as photosynthesis and nitrogen fixation, and of the responses of plants to stress, and to show how this knowledge can be exploited in conventional plant-breeding and genetic engineering. Further, we need greater understanding not only of the ecological underpinnings of integrated pest, disease and nutrient management, but also of the economic and institutional requirements for success. And, he concludes, this is equally true of how we manage our natural resources: rangelands, forests and fisheries. The World Bank (1997) adds to the picture of the challenge to education when it states: we must improve the efficiency of land, water, and chemical use if we are to feed the world's population, expected to exceed eight billion by 2025, without destroying the environment.

Surely this is a challenge worthy of our best efforts and we should be willing to make the educational and institutional changes necessary to meet that challenge. And there is more. Our role must be further defined as the link between the rural and the urban spaces which make up our planet. We have a role in promoting what has become a knowledge intensive sustainable agriculture with inputs for producers and education and information for consumers. We have a goal to educate and train those who protect natural resources and to provide models of effective and efficient natural resources management. Finally, our role as knowledgeable spokespersons for rural development is critical for obtaining government and other support for ensuring that sustainable agriculture can hold its own in a global market-place.

Challenge revisited

Can the agricultural education community ignore the increasing volume of advocacy for change in the way we view our roles; in the way we position ourselves to serve a new sector reality; in the way we join with others in maximising our contribution to society's understanding of agriculture's role in the development process; and in the way we govern ourselves? The question must no longer be whether we espouse change but what type of change and how much do we support? As we have seen from the wide-ranging comments of change advocates, the menu is vast. Too vast for any one institution to tackle alone, too vast for most agricultural education and training systems to fully take on board.. We have to recognise that change is not easy in many systems. Csaki (1999) reminds us that the content of change is not necessarily the same in all regions of the world but is context specific. We also know that the dulling influence of some bureaucracies provides little room in which advocates of change can operate. Hansen (1990) noted three areas where many agricultural universities lacked control:

- Enrolment: where admission policies are controlled by outside government agencies which frequently encourage rapid increases in enrolments without ensuring funding increases to match the expanding numbers
- Programming: where curriculum policy is under the control of a central outside agency, which leaves the university or other education and training institution little if any latitude or incentive for undertaking curriculum innovation.
- Financing: most universities have very little control over the structure of their finances and in many cases the fee structure and faculty salaries are set by an outside agency. Budgetary flexibility is limited and income earned is returned to the Government treasury.

Taking the plunge

We can continue to focus on parts of the AET system and indeed improve quality up to a point. For example much is written and spoken about the virtual university and its potential for linking best practices and practitioners from anywhere in the world to enhance university programs and learning in general. This is a technological application of great potential but it does not answer the immediate question of what is the university, in a system context, going to do about Rural Development. How far will it move from traditional approaches, how far is it willing to partner, to become, in some cases, the junior partner, and how serious will it become about taking the message of agriculture, food and fibre, and sustainable development to society at large? If we are convinced that a change initiative is necessary and that action needs to be taken quickly there are a number of initiatives gaining strength which can provide a valuable support network .

The Global Consortium of Higher Agricultural Education and Research founded in 1998 with the goal of fostering global co-operation for the improvement of higher education and research for agriculture as a prerequisite to solving the food security and environmental problems confronting our world. The consortium aims to serve institutions with programs in agriculture, veterinary medicine, and natural resource management, including the biological, physical and social sciences dimensions of these fields. The consortium founders designed it to be helpful to institutions world-wide that are working to make significant reforms in their systems of higher agricultural education.

The Standing Forum for Discussion on the Integration of Agricultural Education in the Americas sponsored by the Inter-American Institute for Co-operation on Agriculture (IICA) was established in 1999 with a conference at the Organisation of American States (OAS) in Washington D.C. U.S.A. The purpose of the Forum is to help position agriculture and, in particular, agricultural and rural education and training, on the work agenda of political and financial entities and to support modernisation efforts and facilitate integration among institutions and countries.

The Organisation for Economic Co-operation and Development (OECD) through its Directorate for Food / Agriculture and Fisheries held a January 2000 Conference of Directors and Representatives of Agricultural Knowledge Systems (AKS) from 22 OECD countries. The AKS encompasses Agricultural Research, Extension, and Higher Education. The January 2000 Conference noted that most countries see exciting challenges for Agricultural Knowledge Systems (AKS) to contribute strongly to the newly developing societal interests that are wider than traditional agriculture. Mechanisms to encourage, stimulate and reward both institutions and individuals to engage in innovative interactive research, teaching and development work in these new areas still need further development. Many countries, however, identify the limited contribution that AKS has made in recent years to public debate and policy formation as a major weakness which has to be overcome.

The Food and Agriculture Organisation of the United Nations (FAO) has an ongoing program on agricultural and rural education and training particularly concerned with the need for reforms and well reasoned responses to the pressures of change.

The World Bank through the Agricultural Knowledge and Information Systems (AKIS) thematic group in the Rural Family has been working for a revival of interest in agricultural and rural education since 1998. AKIS sponsored an international workshop at the World Bank in late 1999 with the theme: Education for Agriculture and Rural Development: Identifying Strategies for Meeting Future Needs. The Workshop identified a number of Researchable Questions that needed to be answered if the donor community is to make a case for future investments in agricultural education systems.

A number of bilateral donors have been active in supporting innovative agricultural education projects over the past ten years and were identified by Willett (1998).

The Kellogg Foundation support to the US Land Grant Universities in positioning themselves for the future has already been cited as has the parallel Secondary Agricultural Education reinvention exercise.

Key elements in meeting the challenge of change

There are two key elements required to bring about change in agricultural education systems; Vision and Leadership.

Bawden (1998) identifies weak leadership and inappropriate conceptual maps for the development process as a cause of crisis in AET systems and in the broader domain of agriculture and rural affairs. He indicates that where previously the agricultural education and training system (AET) emphasised providing skilled manpower for techno-scientific production agriculture to assure food security, the emerging focus is on developing and promulgating an environmentally sustainable, socially equitable, and ethically defensible agricultural development process that fosters the wellbeing of rural communities and of the biophysical and socio-cultural environments in which they live. To deal with this complex situation Bawden suggests that leaders of AET systems will need to learn how to envision plausible futures (see Box 4).

Box 4: Elements of reform

> Reform in any university anywhere in the world cannot occur unless there is a vision passionately believed in and furthered by leaders. If we want change or reform, it will not happen casually or simply by its bubbling up within a university. There may be ferment for change and a desire for adaptation. But change will not occur unless there are leaders willing to step up and step out and provide direction and articulate a vision that can unite men and women to work for needed change, building on the accomplishments of the university and its history, but pointing unequivocally to the future.

Magrath, 1999

McCalla (1998) asks if the global agricultural science establishment as currently constituted meet the challenge? Possibly, but only if radical changes are initiated now, new partnerships, particularly with the private sector, are initiated soon, and if the agricultural establishment gets out of its isolationist shell and joins the global science community.

We recognise that educational institutions are conservative and slow to change. Paalberg (1998) reminds us that we should not overlook that institutional lag and rhetorical lag serve a useful purpose: providing needed continuity. If we responded fully and quickly to technological change, society would be in disarray. While the pace of change in our institutions and rhetoric has been too slow, some lag does permit accommodation without inducing chaos. Technological changes are the wing feathers, propelling us forward, while institutions and rhetoric are the tail feathers, keeping us on course. Both are needed if we are to fly. This is sage advice but regardless of the pace of change we need to know where we are going and we need the leadership to take us there for a safe landing.

References
Bawden, Richard. Agricultural Education and Training: Future Perspectives. Paper prepared for the Agricultural Knowledge and Information Systems (AKIS) thematic team in the Rural Family of the World Bank. 1998.
Britz, Julian. Higher Agricultural Education in Western Europe. Conference on Globalizing of Agricultural Higher Education and Science. Keiv, Ukraine, 1998. pp.156.
Cleaver, Kevin A. Presentation at World Bank Rural Development Department Retreat. Washington D.C. November, 1999.
Conway, Gordon. The Doubly Green Revolution: Food for all in the 21st Century. Cornell University Press, 1997.
Debertin, David L. There is a Future for the Land Grants, If...Choices. Third Quarter 1992 Published by the American Agricultural Economics Association.

Dlamini, Barnabas M. Public and Private Partnerships. In Proceedings of the Inaugural Conference of the Global Consortium of Higher Education and Research for Agriculture. Iowa State University, 1999.

Engel, Paul G.H. and Wout van den Boer. European Journal of Agricultural Education and Extension, 1995, Vol. 1, Number 4.

Falvey, L. and C. Maguire, The Emerging Role for Agricultural Education in Producing Future Researchers. Journal of International Agricultural and Extension Education, Vol. 4, Number 1, Spring 1997.

Falvey, L., Bright Students for Agriculture: Do we attract them. Is Agricultural Education interesting? Invited paper presented at the Annual Conference of the Australian Institute of Agricultural Science and Technology, Crown Casino, Melbourne, August, 1997.

Falvey, Lindsay. Food Environment Education: Agricultural Education in Natural Resource Management. The Crawford Fund for International Agricultural Research and Institute For International Development Limited. Melbourne 1996.

Hansen, Gary E., Beyond the Neoclassical University; Agricultural Higher Education in the Developing World, An Interpretive Essay. A.I.D. Program Evaluation Report Number 20. Washington D.C. Agency for International Development (IDCA), 1990.

Magrath, C. Peter, Reforming U.S. Higher Education. In Proceedings of the Inaugural Conference of the Global Consortium of Higher Education and Research for Agriculture. Iowa State University, 1999.

McCalla Alex F. Agricultural Education, Science and Modern Technology's Role in Solving the Problems of Global Food Resources in the 21st Century in Conference Proceedings of Globalizing of Agricultural Higher Education and Science: meeting the needs of the 21st century. Kyiv National University of Ukraine. Kiev, September 1998.

OECD Conference of Directors and Representatives of Agricultural Knowledge Systems (AKS). "Summary and evaluation." Paris, November 1999.

Paarlberg, Don. The Land Grant College System in Transition. Choices, Third Quarter 1992. Published by the American Agricultural Economics Association.

Press, Eyal and Jennifer Washburn. The Kept University. The Atlantic Monthly, March 2000.

Reinventing Agricultural Education for the Year 2020. National Council for Agricultural Education, Alexandria, Virginia, USA. 1999.

Ruffio, P. and J. Barloy. Transformations in higher education in agricultural and food sciences in Central and Eastern Europe. European Journal of Agricultural Education and Extension, 1995, Vol. 2, Number 2.

Rural Development, from Vision to Action: A sector Strategy. Environmentally and Socially Sustainable Development Studies and Monographs Series 12. The World Bank, 1997.

Sariego M, Jorge. First Meeting on the Integration of Agricultural and Rural Education in the Americas. Organization of American States, Washington D.C. 1999.

The Economist, June 19, 1999. "Who's afraid?" pp.15-16.

Topel, David G. Partnerships Between Private Sector Agribusiness and Public Higher Education in Agriculture and Rural Development. Conference on Globalizing of Agricultural High Education and Science. Kiev, Ukraine, 1998.

Van Crowder, L. and J. Anderson. Linking research, extension and education: why is the problem so persistent and pervasive? European Journal of Higher Agricultural Education and Extension (1997), Vol.3, Number 4, pp. 241-250.

Van den Boer, W., J.M. Dryden and A.M. Fuller. "Rethinking higher agricultural education in a time of globalisation and rural restructuring." 1995. European Journal of Higher Agricultural Education. Volume 2, Number 3, pp. 29-40.

Wallace, Ian. Agricultural Education at the Crossroads: Present Dilemmas and Possible Options for the Future in Sub-Saharan Africa. International Journal of Educational Development. Vol. 17 No. 1, pp. 27-39, 1997.

Warren, M. Practising what we preach: Managing Agricultural Education in a Changing World. European Journal of Higher Agricultural Education and Extension, 1998, Vol.5, Number 1.

Willett, Anthony. Agricultural Education Review- Support for Agricultural Education in the Bank and by other Donors. Part I: Past and Present Perspectives. Agricultural Knowledge and Information Systems (AKIS) thematic team, Rural Development Department, The World Bank, 1998.

World Bank Assistance to Agricultural Higher Education 1965-1990. Operations Evaluation Department Report No. 10751. The World Bank, Washington D.C. 1992.

From Agriculture Education to Education for Rural Development and Food Security: All for Education and Food For All

Lavinia Gasperini

Senior Agriculture Education Officer, Extension, Education and Communication Service, Food and Agriculture Organization (FAO) of the United Nations, Rome, Italy

Education and Food For All

Ensuring Food for All, and reducing the number of undernourished people to half their present level no later than 2015 as part of an ongoing effort to eradicate hunger in all countries, is the commitment undertaken by the international community during the World Food Summit, convened in Rome in 1996[5].

More than 800 million people do not have access to enough food to meet their basic requirements. Poverty is a major cause of food insecurity and sustainable progress in poverty eradication is critical to improve access to food[6]. More than 1.3 billion people world-wide live in poverty and nearly three fourths of them live in rural areas. Virtually all of them depend directly or indirectly on agriculture for their livelihoods. Despite the continuing process of urbanisation, about 3.2 billion of today's 6 billion world population is rural and this number will be about the same in 30 years time. The total population active in agriculture is about 1.3 billion and this number will not change significantly in the next 10 years[7].

In 1998, the less developed regions as a whole accounted for 97 per cent of the 113 million children not in school. During the same year the number of illiterates was about 880 million[8]. Most commonly, the chronically undernourished are also illiterates and out of school children are a category more at risk of being among the undernourished.

World poverty can be significantly decreased by 2015 if developing and industrialised countries implement their commitments to attack the root causes of poverty.[9] The challenge lies in implementing a common vision for achieving the targets set by the world conferences of the past decade which suggested that we work for sustainable growth that favours the poor and provides more resources for health, education, gender equality, and environmental sustainability world-wide.

The "agriculture-only model of rural development"[10] has proven inadequate to address poverty reduction, rural development and sustainable natural resources management. The latest thinking and good practices in such domains indicate that the empowerment of poor people, policy and institutional reforms in the rural sector leading to participation of

---

[5] Rome Declaration on World Food Security and World Food Summit Plan of Action. World Food Summit, Rome, 13-17 November 1996

[6] Rome Declaration

[7] FAO statistics, 2000

[8] UNESCO, Education for All, Year 2000 Assessment: Statistical document, UNESCO, Paris, 2000.

[9] International Monetary Fund, Organization for Economic Co-operation and Development, United Nations, World Bank Group - 2000 Goals: A better World for All. Progress towards the international development goals, New York, June 2000

[10] The World Bank, Policy and Institutional Reform for Sustainable Rural Development: putting the pieces in place, WBI, Training Course, 2000)

stakeholders needs to be the starting point[11]. Also The Rome Declaration stressed that sustainable development policies should consider education essential for empowering the poor and achieving food security. Research shows that basic education affects small landholders and subsistence farmers productivity immediately and positively, and that a farmer with four years of elementary education is, on average, 8.7 per cent more productive than a farmer with no education. Moreover, farmers with more education get much higher gains in income from the use of new technologies and adjust more rapidly to technological changes[12]. The provision of more and better basic educational services in rural areas such as primary education, literacy and basic skills training can substantially improve productivity and livelihoods.[13] Moreover, many children will be the farmers of tomorrow, and educated children have more chance of becoming more productive farmers. All of the major UN conferences and conventions of the last decade[14], including the United Nations Conference on Environment and Development[15] and the World Food Summit recognised that education and training are indispensable with respect to achieving sustainable development and successfully implementing all chapters of Agenda 21.

In this new millennium, as the global market moves from a technology based to a knowledge based economy (K-Economy), education and training will become even more crucial and access to quality of education will be the yardstick which will differentiate and increase the gap among rich and poor.[16].

---

[11] See for example, International Monetary Fund, Organisation for Economic Co-operation and Development, United Nations, World Bank Group - 2000 Goals: A better World for All. Progress towards the international development goals. New York, June 2000

[12] "The single best measure of basic education impact on economic development, however, is the additional productivity of workers or farmers with more education over those with less. Productivity measures show directly the effect education has on the capacity to produce, and, hence on the potential to increase economic output. A survey done for the World Bank on 18 studies that measure the relationship in low-income countries between farmers' education and their agricultural efficiency (as measured by crop production) concluded that a farmer with four years of elementary education was, on average, 8.7 per cent more productive than a farmer with no education. The survey also found that the effect of education is even greater (13 per cent increase in productivity) where complementary inputs, such as fertiliser, new seeds or farm machinery, are available". Martin Carnoy: The Case for Investing in Basic Education. UNICEF, New York 1992, p. 26, 34 and 41.

[13] "Farmers with little land are highly risk averse, in general, because they have so little flexibility. For them, the difference between a good harvest and a bad one can be the difference between subsistence and hunger. Those small-scale farmers with higher levels of education, however, even with a few years difference in schooling, are better able to adapt innovations to local conditions and therefore more likely to assume risks in changing production techniques." Beatrice Edwards, Rural Education and Communication Technology, paper presented at the First Meeting on the Integration of Agricultural and Rural Education in the Americas; Washington D.C. August 25-27.

[14] Such as the World Summit for Children (1990), the Conference on Environment and Development (1992), the World Conference on Human Rights (1993), the World Conference on Special Needs Education: Access and Quality (1994), the International Conference on Population and Development (1994), the World Summit for Social Development (1995), the Fourth World Conference on Women (1995), the Mid-Term Meeting of the International Consultative Forum on Education for All (1996), the Fifth International Conference on Adult Education (1997), the International Conference on Child Labour (1997) and the Dakar World Forum on Education For All (2000).

[15] Report of the United Nations Conference on Environment and Development, Rio de Janeiro, 3-14 June 1992, vol 1, Resolutions adopted by the Conference. United Nations publications, sales No. E.93.1.8) Resolution 1, annex 2;

[16] In this new millennium the creation and development of a "learning society" in which all children and adults are provided through basic education with the capacity of written and numeric communication, for people to be able to further learn ("trainability" and "learnablity") and make improvements in their own lives and sustain development in the information age is especially important in poor countries where relatively few jobs are available and development has to come from widespread ability in the population to improve their livelihoods. In Martin Carnoy: The Case for Investing in Basic Education. UNICEF, New York 1992

Why International Assistance to Agriculture Education has declined?

As "the agriculture-only model" has proven inadequate to address rural development, so has agriculture education and training (AET). Education and training needs to address rural development, sustainable natural resources management and poverty reduction, with a broad, holistic focus by redefining its strategies and responsibilities and expanding its target.

We can identify two series of reasons responsible for the crisis of AET:
(a) reasons specific to strategies and targets of the AET system, (b) and reasons common to the crisis of the "business as usual" education and training paradigm prevailing until the eighties.[17]

(b) The specific reasons for the AET system crisis have been analysed on many occasions, among others, by Maguire[18], Lindley[19], or by Willet [20]: agriculture education and training have been isolated from the market place and from the rest of the education system. This isolation has been leading to curricula irrelevance, falling teaching and learning standards, unemployment of graduates and, thus, decreasing investment support. Responses to such a crisis were mostly fragmented, and inward looking, lacking a vision and a systemic approach. Operating in a sort of "ghetto", or an "Ivory tower", according to the situation, AET has taken responsibility only for a reduced clientele, including the students of vocational education and training institutions, and of the higher agricultural education, and has not addressed the needs of the vast majority of the rural population, who represent a great percentage of those 800 million undernourished and illiterates. This has become, instead, the priority target for poverty reduction and education strategies in the decade of the nineties. By disregarding the educational needs of vast numbers of the rural population, donors and governments have built an agricultural human resource pyramid which could be considered almost as an inverted pyramid, where the absence of a diffuse general nor specialised knowledge, can limit national efforts to implement sustainable policies for agriculture, rural development, food security and poverty reduction efforts.

(c) Some of the reasons for the crisis of AET are common to the crisis of the "business as usual" education and training paradigm[21]. As in other domains of international assistance to education, different prevailing paradigms[22] could be identified over time also for AET.

---

[17] Kenneth King, Aid and Education in the Developing World. London. Longman, 1991; or Introduction: new challenges to international development co-operation in education. In Changing International Aid to Education: Global patterns and national context

[18] For examples: Charles Maguire, Education for Agriculture and Rural Development: Identifying Strategies for Meeting Future Needs, powerpoint slides, presented and the World Bank AKIS Workshop, Washington, 1-3 December 1999; First meeting on the Integration of Agricultural and Rural education in the Americas, Washington DC, August 25-27 1999; From Agriculture to Rural Development: critical choices for agricultural education, paper presented at the 5th European Conference on Higher Agricultural Education, September 10-13, 2000

[19] William Lindley, Quality Improvement in Undergraduate Education, Proceedings of the Inaugural Conference of the Global Consortium of Higher Education and Research for Agriculture, July 22-24, 1999, Amsterdam, The Netherlands

[20] Anthony Willett, Agriculture Education Review, Support for Agriculture Education in the Bank and by other Donor. Part I . Past and Present Perspective. AKIS thematic team, The World Bank Rural Development Network, October 15, 1998

[21] Kenneth King, Aid and Education in the Developing World. London. Longman 1991; or Introduction: new challenges to international development co-operation in education. In Changing International Aid to Education: Global patterns and national context. UNESCO/NORRAG, Paris 1999.

[22] As suggested by Stephen Heyneman in: Development aid in education, a personal view, in: Changing International Aid to education: Global patterns and national contexts".Pages 132-146

During the period (1960-80), the dominant rationale was education and training for **economic growth**. International assistance was funding mostly public sector oriented, donor-driven "enclave projects" of vocational and technical training or higher education, where the public sector was the deliverer of education and training, and often also the main expected employer of graduates. During this period local authorities ownership or involvement in policy formulation and strategic decision making was quite limited. Investments in primary education were a minor percentage of total international aid to education: in 1981-1983 about 7.4 percent of direct aid to education, versus 39 percent for secondary education (general, training of trainers and technical) and 34 percent for higher education[23]. Higher education was regarded by donors as a politically more rewarding investment. Investments were predominantly supply driven and focused mainly on hardware (equipment, vehicles and construction), international experts and overseas training. Recurrent expenditures that would allow project sustainability and institutional capacity building were disregarded. Salaries of overseas experts would absorb about 44 percent of the direct aid to education, and scholarships abroad about 17 percent[24]. In this context aid to AET education and training was aimed at training the "right number" of individuals, calculated by manpower forecasting techniques, equipped with the training required by factories, farms and companies, in order to allow them to deliver products and services which would allow their economies to grow. As Bawden recalls "Growth was the aim and the objective of education and training assistance was to provide manpower to support techno-scientific production and productivity growth. Little if any concern was expressed for rural development and for possible long term impact on biophysical and socio-cultural environmental aspects of development"[25]. Although physical and staff development targets were met, sustainability became a serious issue since the recurrent expenditures of new educational facilities grew beyond institutional economic capability.[26]

By the end of the seventies and beginning of the eighties the international debate started questioning previous priorities and strategies of international aid to education and training, as well as the manpower planning techniques[27]. The focus started shifting from education for economic growth to education for development as an integrated complex social, cultural and economic process[28] that should be aiming at contributing to poverty reduction.

The UNDP Human Development Report, in 1990, set the decade's development agenda, emphasizing the shift from the development paradigm that narrowly focused on growth as measured by Gross Domestic Product (GDP), to Human Development, measured by the Human Development Index, a composite index based on three main indicators: longevity, educational attainment and standards of living[29]. The education sector agenda was set in 1990

---

[23] The World Bank Education in Sub-Saharan Africa, The World Bank, Washington DC1988. Pag 108-9 (French version)
[24] Education in Sub-Saharan Africa, Washington, quoted
[25] Richard Bawden, Agriculture Education Review. Part II. Future Perspectives. AKIS thematic team, The World Bank Rural Development Network, November 12, 1998
[26] Willett, op. cit.
[27] "Dans les pays industrialisés le problème ne consistait plus à faire face aux besoins en main d'oeuvre, mais au contraire à affronter le chômage.(...) De leur côté, les pays en développement rencontraient des difficultés dans leur exercice de prévision des relations formation-emploi, du fait du manque de données et de moyens, et s'apercevaient des obstacles politiques que soulevait une telle planification. Pour toutes ces raisons l'approche main d'oeuvre comme instrument de prévision et de planification a été un peu partout abandonnée, in: Olivier Bertrand, Planification des ressources humaines: méthodes, expériences, pratiques. UNESCO/IIPE, Paris, 1992
[28] A milestone in the international educational debate has been the publication of the World Bank 1988 book on Education in Sub Saharan Africa (quoted above).
[29] "The Human Development Index is a composite index based on three main indicators: longevity, as measured by life expectancy at birth; educational attainment, as measured by a combination of adult literacy (two-thirds

by The World Conference on Education for All (EFA)[30], which stressed the need to invest, as sector first priority, on basic education, formal and non formal, including early childhood, primary education, literacy, and basic population, health and agriculture skills for life. Although vocational and technical training[31] and higher education were acknowledged as important components of education systems and of international aid, priority was given to EFA. This meant less funding available for traditional intermediate and higher education projects also in the agriculture sector, especially if formulated in isolation from a wide education sector approach, a systemic policy, or a clear focus on poverty reduction. While the interest for agriculture education projects focused on limited stakeholders involvement declined, a new concern for the needs of the disadvantaged, the poor, and for the unequal educational opportunities of rural population has emerged.

Summarising, in general, and with minor differences, donors' priorities over the last two decades shifted from vocational and technical and higher education to basic education; from isolated project to coordination and "sector wide approach"[32], aiming at facilitating local ownership of national policies and programs. Sustainability, institutional and system reform, definition of long term sector policies, diversification[33] and a strong concern for sustainability, relevance, responsiveness and efficiency became today's imperatives for investments in education and training systems. Within such a context awareness has started developing[34] about the fact that although education is universally acknowledged as a prerequisite to build a food secure world, reduce poverty and conserve and enhance natural resources, educational opportunities are not equally distributed [35].

---

weight) and the combined gross primary, secondary and tertiary enrolment ration (one third weight); and standards of living, as measured by GDP". In UNDP: Human Development Report 2000, UNDP, Oxford University Press, New York, 2000

[30] World Conference on Education for All, (Jomtien, Thailand, 5-9 March 1990). World Declaration on Education for All and Framework for Action in Meeting Basic Learning Needs. New York, Inter Agency Commission for WCEFA UNESCO, 1990

[31] See, on the topic, for example, Groupe de travail pour la coopération internationale en matière de développement des compétences professionnelles et techniques: Politiques des agences en matière de développement des compétences professionnelles et techniques. Un résumé des réunions de Francfort (Novembre 1996) et de Londres (Mai 1997). Une publication conjointe du DDC, BIT, and NORRAG, Berne 1997. Or: The World Bank, Vocational and Technical Education and Training, A World Bank Policy Paper, Washington DC, 1991. Working Group for International Cooperation in Vocational Skills Development: Donor Policies in Skills Development. Reforming education and Training Policies and Systems, Geneva, April 1998 Working Group for International Cooperation in Vocational Skills development: Debate in Skills Development. Sector Program Support and Human & Institutional Development in Skills Development, Copenhaghen, June 1999.

[32] for "Sector Wide Approach " see, for example: Lars Rylander and Martin Schmidt: SWAP Management, Experiences and Emerging Practices", Sida, Stockholm, 2000

[33] meaning by this a new role for the private sector

[34] Proceedings from the Workshop: Education for Agriculture and Rural Development: Identifying Strategies for Meeting Future Needs", Washington The World Bank, December 1999, and: Principles for Developing an FAO Strategy in Support of Agricultural Education and Training in: Issues and opportunities for agricultural education and training in the 1990s and beyond, agriculture education group, SDRE, FAO, Rome 1997, p.69.

[35] "Due to increasing urbanisation, fed by out migration from rural areas, governments often give priority to urban needs for health, education and social services, at the expense of rural areas and the agriculture sector. Such trend leads to even greater impoverishment in rural areas and leads to higher levels of migration.(...). There is thus a need for comprehensive rural development policies, that protect rural population from further marginalisation by a more organized urban sector with a greater political voice. In: "Issues and opportunities for agricultural education and training in the 1990s and beyond, agriculture education group, SDRE, FAO, Rome 1997, p.17

- Access to education is lower among rural children, youth and adults: The gap between urban and rural illiteracy is not closing; in several countries, rural illiteracy is two or three times higher than urban illiteracy.
- Quality of education is lower in rural areas: curricula and textbooks in primary and secondary schools are often urban biased, irrelevant to the needs of rural people, and seldom focus on issues such as skills for life and rural development.
- Institutional capacity to address education for rural development and food security needs strengthening: ministries of education and their universities, ministries of agriculture, health, finance etc., often lack awareness and coordination in targeting the needs of the poor, sustainable rural development and food security.

So how do we move forward?

Given the multifaceted character of poverty and food security, the FAO Education Group sees the following areas for systemic action:

1. Targeting multiple stakeholders[36], focusing on "Education for All" and Food for All

We want to move from production agricultural education, to a systemic and inclusive approach embracing a wide range and large numbers of stakeholders through formal and non formal education, at all levels of the education system. We intend to contribute to granting access to quality education to the absolute poor, rural landless, urban dwellers, women, members of minority groups and other disadvantaged groups, especially those in isolated backward areas and to all food insecure groups. While moving from the traditional AET approach to Education for Rural Development and Food Security, (EFRDFS) we shall still place great emphasis on vocational and technical training and higher education for agriculture and rural development, while focusing, with priority, on satisfaction of basic learning needs of rural populations. This is what we call "Education and Food for All".

2. Contributing to placing education at the core of the global and national development agenda [37] and food security agenda, by focusing on the following priorities:

❖ Expanding access to education and improving school attendance in rural areas by promoting or supporting:
- Initiatives aiming at improving children's health, providing food for students, easing the financial burden on parents who usually have to feed their children, and in some cases generating income for the school, such as school canteens and school gardens, fish ponds and raising of animals.
- The use of information and communication technology, and distance education.
- Education of rural girls and women.
- Life long education and skills for life in a rural environment.

❖ Improving the quality of education for rural development and food security by supporting :
- Participatory curriculum development, and teacher training to respond to rural development needs and farmers' demands, at all levels of the education system[38].

---

[36] Maguire.
[37] see James D. Wolfensohn, President, The World Bank: Placing Education at the Core of Development, Presentation at the World Education Forum, Dakar, Senegal; April 27, 2000
[38] Alan Rogers, Peter Taylor, Participatory Curriculum Development in Agriculural Education. FAO, Rome, 1998;

- Environmental education, across the curriculum[39], and awareness raising in relation to sustainability of current patterns of consumption.
- Nutrition education [40], including school gardening and small animal care which bring alive the content of science and social studies, provides shared experiences for language development and settings for mathematics. These can also provide life skills, basic entrepreneurial and self employment skills, while also contributing to enhance the relevance of the curriculum and quality of education [41].
- Basic financial literacy for children and adults (through, for example, schools saving clubs) and marketing, and rural financial management education in intermediate and higher education.
- HIV/AIDS prevention to address its impact on agriculture and rural livelihoods.
- Agricultural universities and vocational training institutions to improve their role of service to farmers, rural children, youth, and adults, and their interaction with basic and intermediate education.

❖ **Strengthening Institutional Capacity** in planning and managing education for rural development and food security. Efforts to ensure Food for All need to be closely co-ordinated with those aiming to reach Education for All, since the results of both programs are interdependent. Targeting rural population educational needs requires an increased partnership and an interdisciplinary approach among government entities (Ministries of Education and their Universities, Ministries of Agriculture, Health, etc), and with private national and international organizations, civil society, mass media, and religious organizations. We intend to contribute to:

- The definition of a systemic approach to education for rural development and food security, addressing all levels of education, with priority on basic education (primary formal and non formal education, adult literacy and adult education and basic skills for life).
- Research and dissemination of best practices and case studies which illustrate the contribution of education to sustainable agriculture and rural development and food security.
- Training of policy makers and managers on education for rural development and food security

Last but not least we attribute great importance to:

3. Fostering Interdisciplinarity and new partnership, which are two basic principles underlining FAO strategies and objectives[42]. The international conferences and summits convened in the nineties indicate the need for a concerted attack on poverty and environmental degradation. The new paradigms emerged in such fora indicate that new

---

[39] See for example: F.M. Schlegel, Ecology and Rural Education, Manual for Rural Teacher, FAO, Rome 1995; Gagliardi & Alfthan, Enviromental Training, ILO Geneva 1994; Intégration des thèmes de l 'environnement et du développement durable dans les programmes d'éducation et de vulgarisation agricoles, Consultation d'experts, FAO, 1993. Sylvia A. Ware, Science and Environment Education Wiews from Developing Countries, The World Bank Washington D.C. 1999

[40] See also Beryl Levinger, Nutrition, Health and Education for All. UNDP and EDC, Ewton, Massachusset, 1994

[41] Peter Taylor and Abigail Mulhal, Contextualising teaching and learning in rural primary schools: using agricultural experience. Vol 1 and 2, Department for International Development, serial number 20, London 1997

[42] The Strategic framework for FAO, 2000-2015 Paragraph 31. FAO, Rome 1999

alliances need to be adopted in order to address development, and the efforts to ensuring Food for All need to be closely coordinated, among others, with those aim of reaching Education for All. These new partnerships for education for rural development are needed at global and regional level (among UN and other development agencies and organisations) and at national level, between governments (Ministries of Agriculture, Education, including the Academia, Ministries of Health, Finance etc.), local government bodies private and public organizations, universities, civil society, mass media.

It is time to make decision for priorities for change. There is much to be done, we need to run against time to go beyond the set goal of halving the number of hungry and illiterate people by the year 2015, and ensuring that all children, particularly girls, are in school. These are not maximum goals, but minimum goals. New partnerships mean working together. Dear colleagues, if we are working together, breaking ancient walls and bridging our efforts, we can make it. Together, we can build a better world for all.

Implementing change: challenges for institutional structures and management

Cees Karssen
Wageningen University and Research Centre, The Netherlands
President, Interuniversity Conference for Agricultural and Related Sciences

The challenges

The central theme of this 5th Conference on Higher Education in Agriculture: *From production agriculture to rural development* describes undoubtedly an important new challenge for the institutions of higher education in agricultural and related sciences. Previous lectures have demonstrated that in great detail.

As has been stated before integrated rural development throws up exciting challenges for agricultural education. Warren(1999) states correctly that our institutions *"have a mission to help rural society grow and develop, and to provide the new skills and new knowledge needed. Who better to do this than institutions which for many decades have been so closely wedded to rural life - institutions with a long history of interdisciplinary operation; with substantial knowledge and contact networks in rural society; trusted by a large part of rural society on the basis of past achievements in the agricultural arena?"*.

It is my task to suggest ways in which this major change can be implemented into the structures and the management of our institutions. In the performance of my task I need to go somewhat beyond the borders of the central theme of the Conference. In my view food production is still the main task of agriculture and thus still has to be taught in our institutions. It is one of the biggest challenges of our generation to produce sufficient, healthy food for the still growing world population. This is not to say that rural development is not important. However, the theme of the Conference seems to suggest that the attention for rural development has to replace that for agricultural production. I suggest therefore as a theme *"Agricultural production in harmony with rural development"*.

Also in the area of agricultural production major changes have to be implemented. I mention a few:

* For a long time the priority of agricultural production has been on the increase of productivity. However, during the last decades the main accent shifted from *quantity* to *quality and, recently, to food safety*.

* To keep in touch with this development our institutions have to cover much more aspects of the whole *integrated food chain* than before. The chain begins with the isolation or construction of sufficient genetic variation, followed by breeding, crop production, technology and nutrition. The chain may be either directed to food, fodder or non-food products.

* The control of the chain shifts from *supply-driven to demand-driven*. The decisions in the chain are no longer exclusively in the hands of the producers (farmers, co-operatives), the market (consumers, retailers) plays an increasingly dominant role.

* Another important development is that the market shifts from *the local market to the global market*.

Agricultural research and education have become increasingly *science-driven*. The traditional basic natural sciences (chemistry, physics etc) were involved already for a long time. Modern times are characterised by an increasing influence of *Life Sciences* and *Ecological Sciences*. Both multidisciplinary fields have a growing impact on the modernisation of agricultural production into the direction of *Agrobiotechnology and Organic Agriculture*. So it is quite clear that education and research in the traditional institutions for agricultural sciences are in a process of major change. The institutions have to widen and to modernise their scientific field their area of interest and therefore have to integrate new disciplines into the multidisciplinary family that is already present. As a consequence new curricula have to be designed and new staff to be selected and appointed and , bitter consequence, the volume of staff in other parts has to be reduced. Strong managerial skills are needed. Certainly, because scientists are by nature antagonistic any change in the organisation of their working environment.

The conditions for implementation

Two conditions have a major impact on the implementation of change.

Changes in the position of agriculture in society. Agriculture is somehow the victim of its own success, a success that allows us to produce food, fodder and raw products with less people, less land, less input and less water. Nevertheless, the position of agriculture in society deteriorates. It has been damaged by several environmental side effects of agricultural production, like over-use of fertilisers, dung problems, and misuse of pesticides. Moreover, consumers are concerned about the quality of food products and the way they are produced. Also that trend has a negative effect on the level of acceptance of agriculture by society. Risk perception and risk acceptance need explicit attention. Agriculture has to renew its licence to produce in order to remain an accepted and highly valued industry. In general, agriculture has lost part of its dominant position on the political agenda of Europe and of different European countries. As a consequence diminishing governmental financial support is reported in a number of countries (Documents of the OECD conference on Agricultural Knowledge Systems addressing food safety and environmental stress, Paris, 2000).

Changes in the interests of students. Although agricultural universities and faculties have often been at the forefront in the demand for change in agricultural production, the lower impact of agriculture in society and politics has also affected them in a negative way. The interests of students in undergraduate studies decreased considerably in many industrialised countries over the recent years. This phenomenon is certainly partly due to the greater appeal of many other studies, but it is also due to a misinterpretation of agricultural sciences. Modern trends in agricultural research and education often take considerable time to reach the world outside the universities. It is a weak consolation that the phenomenon is not restricted to agricultural higher education but also occurs quite seriously in a number of technical and natural sciences like physics and chemistry. The lower intake of students in agricultural production studies is in many institutions more or less neutralised by an increase in modern studies like food sciences, nutrition, life sciences and above all environmental sciences. It is of help that the political interest in integrated environmental development is growing, also on a European scale. Protection of the environment and conservation of natural features and appearance of the landscape is a new societal priority and attracts students.

The institutional structures

The structure of the institutions for higher education in agricultural and related sciences in Europe is rather diverse. Roughly speaking two models can be distinguished: (1) *separate universities* like for instance the Swedish Agricultural University (Uppsala), the National Agricultural University of the Ukraine (Kiev) and Wageningen University, and (2) *special faculties within larger general universities*, this model is found in for instance Belgium, the

UK and Germany. In countries where higher education occurs at two levels (academic and non-academic whereby the latter is often restricted to undergraduate studies) a similar distinction exists.

Also in the *disciplinary field* that is covered by the institutions large differences exist. Some faculties still restrict themselves to agriculture *in sensu stricto*. However, most institutions have already diversified their disciplinary field to a certain extent, by broadening their courses in the food chain and by developing environmental sciences. Interestingly, forestry and veterinary sciences are sometimes organised in separate institutions, and are sometimes joined with agricultural sciences.

These differences in content are in several institutions already expressed in *the names of the institutions*. An increasing number of institutions avoid agricultural or use it in combination with other names. New names are for instance *Faculty of Agricultural and Applied Biological Sciences* (Gent and Leuven) and *Faculty of Life Sciences* (Weihenstephan). My own university last year made the choice for the 'neutral' name Wageningen University instead of Wageningen Agricultural University, the same happened in Hungary with the merger of some universities and faculties into the Saint Stephan University (Gödöllö).

The number of institutions differs strongly between countries. In some countries all higher education in agricultural and related sciences is restricted to one university (*e.g.* Sweden, Norway); in other countries it is subdivided over many small universities and faculties (*e.g.* Poland). This situation is one of the reasons that also the size of the individual institutions strongly differs: from a few hundred students to 4-5000. In general, the institutions for higher education in agricultural and related sciences are among the smallest universities and faculties in the field of higher education.

An urgent need for implementation of changes

I have indicated in the above sections that the institutes for higher education in agricultural and related sciences have to implement two major changes. (1)They have to *diversify their education and research programmes* by including a wider range of food chain elements and by a strong stimulation of environmental issues. (2) They also have to *modernise their programmes* by the introduction and development of modern sciences like life sciences and agro- and landscape ecology.

The implementation of those changes is *not optional, it is a must*. Without these changes our institutes might be dangerously shifting to the periphery of the scientific spectrum. There might also be a serious threat that the scientific and educational interest for the problems of food security, environmental protection and rural development will be scattered over a number of faculties or institutes. I do not favour such a development. I believe that the integrity of agricultural and related sciences has to be maintained. Progress depends on a strong interdisciplinary approach by a large family of disciplines. Organisational and physical separation between disciplines will harm the fulfilment of our mission.

Organisational weaknesses.

Implementation of the changes is far from simple. It is hindered by a number of organisational weaknesses. Diversification and modernisation need substantial amounts of money which often are not available due to *a general weakness of the financial position* and to the lack of political support for new investment. Another, related constraint for the implementation of change is, as I mentioned above, *the small size of most institutions*. Already without the need for change small universities and faculties have problems to maintain a sufficient critical mass in their disciplines. Diversification is then out of the question. Internal shifts will be certainly of some help: a reduction in for instance agricultural production sciences will offer the possibility to introduce a new discipline in environmental sciences. However, I am afraid that stronger measures are needed to implement change.

Options for implementation: need for partners

Diversification and modernisation ask for more co-operation. Most faculties and also most independent, specialised universities need partners to fulfil the needs for change.
But not every partner is acceptable. Partners have to show: (1) a willingness to assist in the *realisation of our central mission* ("Helping people to acquire sufficient, healthy food, in a vital world") and (2) a readiness to *divide tasks, disciplines, chairs and programmes* with us.

From my own experience I know that the road to co-operation is often difficult. The development of partnership has to obey the rule 'structure follows strategy'. A discussion about the structure of the co-operation in an early stage nearly always obstructs the partnership. Nevertheless, in the end, every partnership needs some structure. An upwards moving succession of structural changes may begin with the integration of programmes and courses and may end, via sharing of facilities, and merging of Executive Boards, with a full merger of the institutions. However, it might also be preferred that co-operation retains a virtual character.

Local, national or international partnership

Co-operation can occur at local, regional, national or international level. The shorter distances between *local and regional partners* may offer unique options for the physical sharing of facilities and for the regular exchange of teachers and students in the framework of joint programmes. A merger might be then a perspective. In many countries such partnership will help to reduce the number of institutions and therefore will help to increase the critical mass in several disciplines.

In smaller countries, like the Netherlands, the same options will be open at *national level*. The co-operation that is in progress within *Wageningen University and Research Centre* is an example of the concentration of all national academic education and research in one Centre. The co-operation started in 1997 with the merger of the Executive Boards of the Wageningen University (then still named Wageningen Agricultural University) and the Research Institutes (DLO) of the Ministry of Agriculture, Nature Management and Fisheries. The institutes are for the main part also housed at Wageningen. They perform strategic research for public and private sources. At the beginning of 2001 the Institute for Applied Research will join the co-operation.

At the end of the process a Centre will be realised with a staff (academic and non-academic) of about 7000 persons that covers the whole knowledge chain from fundamental to applied research and academic education. We have taken good care that the specific character of the university and the research institutes has not been lost during the integration process. It is of help that both partners are still jurisdictionally independent, also the money streams are not intermingled yet. In practice this separation does not hinder the integration of programmes, the development of joint policies towards the market, the politics, the globalisation of the Centre and so on. A foremost reason for the co-operation is that together the partners in Wageningen UR can maintain the coverage of a wide field of disciplines and themes. Nevertheless, Wageningen University also needed more strength in the field of fundamental, basic sciences. Therefore we seek increased co-operation with two neighbouring universities, Utrecht University and the Catholic University Nijmegen, in common fields like genomics, food and health, biology and environmental sciences.

The Netherlands have the advantage of small size, but I can assure you that the psychological barriers against co-operation are as strong as everywhere. Or, in other words, such an integration process must never be underestimated. It is hard labour and it needs a strong belief by the management and it asks for a strong flexibility among the staff. In our case it has been of great help that the co-operation was initiated by the central Government and the Dutch Parliament. In bigger countries such a concentration in one institution might be impossible, but a few of such concentrations will be better than too many and too small institutions scattered over the country(side).

A good example of *international partnership* is the co-operation of the Scandinavian and Baltic Agricultural Universities in *the Nordic University*. The partners seek options for divisions of tasks between the different national universities. The ERASMUS, TEMPUS and SOCRATES programmes of the Europen Commission also highly stimulated co-operation in Europe between institutions.

A role for European networks and organisations

It may be concluded from the previous section that national and international co-operation is important to implement change. Partnership is an important means to join forces and to maintain critical mass in certain disciplines. European networking may facilitate co-operation between institutions. A relevant network is ICA (the Interuniversity Conference for Agricultural and Related Sciences in Europe). ICA is an association of European universities and faculties in our field. It aims to associate the institutions by means of the rectors and deans in order to develop a common European policy towards the future of our institutions and to stimulate networking. Seventy institutions are members, more have to follow. ICA also offers an umbrella to the different disciplinary or other networks like the Forestry network SILVA, the network of Agronomists EUROCROP, a network of international relation officers IROICA and, hopefully, so on. The Committee that organises these Conferences on Higher Agricultural Education will, hopefully, agree that the conferences fit perfectly in that scheme. Networks will not lose their independence, but will be stimulated to work together as optimally as possible with the other groups under the umbrella. It is again a question of joining forces.

ICA sponsors AFANet, an EU SOCRATES Thematic Network, which aims to stimulate the development of a European dimension to education and co-operation in universities, faculties and colleges in Europe, offering degree programmes in agriculture, forestry, aquaculture and the environment. The EU SOCRATES Thematic Network Project should lead to curriculum development or other outcomes which will have a lasting and widespread impact across a range of institutions. Currently some 100 institutions are members of AFANet. In SOCRATES II a further 100 hope to join. For further information see the Web site at http://www.clues.abdn.ac.uk:8080/demeter .

Some selected outcomes of AFANet demonstrate the range of approaches that have been undertaken so far and their impact on a European dimension. By means of an international workshop, a group of colleagues worked and tried to set out to bridge the conceptualisation of issues of sustainability with the integration of these issues in the curricula of higher education. A book is the result. A SILVA group developed course modules on Forestry in changing societies in Europe. Another group prepared and delivered a professional development course for English language teachers. The IROICA group produced a manual of Good Practice for international relation officers. Through a 'grass roots' approach, AFANet helps to implement the attitude for change among members of staff of our institutions. An essential step in the implementation process.

Agricultural Education in transition – facing up to the issues

Contributed papers

Education and Professionalisation in English Agriculture

Brassley, P.

Seale-Hayne Faculty, University of Plymouth, Newton Abbot, Devon, TQ12 6NQ, UK

Between 1918 and 1925, according to Howkins (1991: 281), the focus of power and deference in rural England shifted from the landowner to the farmer. To be more specific, we might argue that it shifted to the bigger farmers, of whom there were a greater proportion in England than in continental Europe. This paper examines the contention that the result of the social and political hegemony of this group was a greater degree of professionalisation in British agriculture than in other western European countries, and that this process had profound consequences for the development of agriculture and rural society in England. The terms professional and professionalisation are used here in the same way that they are used by Perkin (1989), although it should be noted that Perkin does not argue for the professionalisation of agriculture.

Changes in agricultural education were both a cause and an effect of this process. Nineteenth-century educational initiatives reflected the dominance of the great landowners in England. This dominance disappeared in the first half of the twentieth century, and the pattern of agricultural education which then emerged was more determined by the priorities of the large-scale commercial farmers who took over as the politically and socially dominant rural class. They distinguished themselves from the smaller farmers and farm workers, and achieved control over numerous rural institutions, from agricultural shows and breed societies to the county executive committees which administered wartime agriculture, the National Farmers' Union, and the marketing boards which existed between the 1930s and the 1990s. The agricultural education system was organised from its inception along class lines, and so changed with them, and the early agricultural research and extension services were seminally influenced by the pioneering professionals. It can also be argued that there was little movement between the class boundaries: few of the small farmers or farm workers moved into the emerging professional class.

This class-based model suggests that the hegemony of the professional farmer class accounts for many of the present institutional features of English agriculture, such as the weakness of the co-operative movement, the political power of the NFU, and the changes in farm size structure over the twentieth century.

References

Howkins, A. (1991) Reshaping Rural England: a Social History 1850-1925   London: HarperCollins

Perkin, H. (1989) The Rise of Professional Society: England since 1880 London: Routledge

To survive the change or to cope with it? (Education for rural development)

Lošt'ák, M.

Dept. of Humanities, Faculty of Economics and Management, Czech University of Agriculture in Prague, 6 – Suchdol, Czechia.

Rural areas are not static entities. They live under omnipresent change generated by the essence of modern society but they also contribute to the generation of this change as well. Similarly, the issue of sustainable rural development cannot be understood without its being framed within the concept of the social change. This paper attempts to outline the actions of local rural actors (represented by former communist collective farms and also by newly emerging individual private farmers) to cope with the changes generated in Czechia at the national (and international) level after 1989. To achieve this goal, the paper also considers the influence of various types of education (and related skills) to manage the changes generated by rural development.

The paper investigates two types of change from the planned economy (equities) and *nomenklatura* farm management (hierarchy) of prior to 1989: (1) changes controlled and managed by their local actors (labelled as transformations which had foreseeable outcomes and were guided by certain world-view, allowing actors to cope with the *anomie* of the change), and (2) spontaneous and yet fundamental changes which were not under the actor's control (labelled as transitions having, at least at the beginning, no foreseeable outcome; their actors experienced a strong *anomie* in the process of change). Coping with changes depends, *inter alia* on actors' social and cultural capital whose building is one of the major tasks of higher education. The paper uses the case of one village with its farms (both former collective communist and newly starting individual private farms) to show how the internal order of the collective farms was re-created and how the external order transcending the borders of the farms was changed in the case of both the former communist and newly emerging individual farms. The paper thus points out how the actors with certain educational background coped with the change after 1989 inside their farms (farm management hierarchy) and outside their farm (equity relations with trading partners) especially with regard to the sustainability of the locality in which they operate. Based on these empirical data, an author attempts to argue for an education furnishing the graduates not only with technical skills but also with social competencies to be able to active manage and to cope with the social change, instead of mere passive surviving this change.

An Interdisciplinary Approach to Addressing Sustainability, Issues in the Agri-Food Sector - "Agricultural Education in Transition: Adapting Curriculum to New Demands"

Stonehouse, D.P.

Department of Agricultural Economics and Business, University of Guelph, Guelph, Ontario, Canada.

There is a need to inculcate an appreciation of sustainability concepts, issues and objectives in the minds of the next generations of landowners, not only in Europe, but globally. There is a lingering feeling that farming is still to be viewed as a way of life, despite on-going technological advances that would make farming more of a science. Such a feeling is reinforced by the view that agriculture is more than just food production; it involves efficient use of scarce resources; it involves stewardship of the natural resource base, the environment, and the ecological balance of species; it involves sustaining rural communities; it involves maintaining the aesthetic appeal of the countryside. Sustainability then, as implied by all these viewpoints can mean different things to different people. Nor is any one of these viewpoints necessarily to be considered the "correct" one.

How then should we proceed to educate future landowners in all these matters? It is posited that one ought to begin by defining the term sustainability. Because sustainability is likely to be interpreted quite differently by members of distinct disciplines engaged in higher agricultural education, a definition of sustainability should be offered from the perspective of each of those disciplines. Reconciliation of these differences and methods of accomplishing this should be a necessary part of the education process.

It is further posited that a deeper and more accurate meaning of sustainability in an agri-food sector context is likely to be best obtained by the student through a team-teaching approach. Composition of the team should ideally encompass all disciplines, from the bio-physical sciences to the socio-economic sciences, plus engineering and medicine. A case study approach is advocated as one effective means of teaching an appreciation of and need for a balanced interpretation of sustainability. The case study setting affords adequate opportunity for evaluating sustainability issues in specific economic, environmental, biological, sociological, or other disciplinary contextual circumstances.

Integrating CEE countries into the EU:
A challenge for rural development research

Kancs, A.

Institute of Agricultural Development CEE, Halle/Saale, Germany

Introduction: Research questions arising from the EU-enlargement

Most of the Central and Eastern European (CEE) applicant countries lag still far behind the EU member states in terms of socio-economic development (European Commission 1997). On the other hand their current institutional capacity is not able to manage the Structural Funds which will be given in a framework of pre-accession assistance, though as stressed by the Agenda 2000 structural policies will be the "second column" of the future EU-economic development policies in the future. Hence many of the structural policy options aim at the development of rural areas, they are of major importance for most of the current member states as well as of the applicant countries (European Commission 1997). The acquierement of structural policy measures in the framework of EU enlargement raises many questions by the current and future members most urgent of which are discussed briefly below.

The first one is the question of defining regions for structural assistance and policy instruments to be applied in these regions of applicant countries in the pre-accession period (Nuppenau, Thiele 1997). The second one is caused by the fact that e.g. the cohesion funding for poorer present members has to be cut down in favour of accession countries when budgets have to be kept stable, and that the overall structural adjustment funds will also have to be redistributed. An open question herein is how competition effects between EU-members and applicants have to be dealt with in terms of budget allocation and political decision making processes (World Bank 2000). A further question to be answered is whether there could be a more differentiated set of accession options than the existing, which only discriminates full accession or the trade union option without acquiring any common structural or agricultural policy option. In order to be able to answer these and similar questions one should know exactly the state of the socio-economic development in applicant countries and their regions. Rural development research as a main rural development contributor has to take a leader role providing knowledge for political decision makers for solving the questions mentioned above. Moreover, this research has to cope with specific factors and characteristics in rural development, namely regional disparities and regional variability (Espotti, Sotte 1999). The major concern in our research is to further develop the knowledge-base of the economic situation in the different CEE countries and in their regions, to examine the development of regional policy as well as to derive tasks for further rural development research.

Methodology: Statistical multi-regression analysis

In this research, comparisons between CEE accession countries and current member countries and their regions are made in terms of PPS (purchasing power standards).

Results: Increasing regional disparities in the union

While enlargement provides, above all, the opportunity for maintaining stability and improving prospects for growth in Europe, there is little doubt that it presents a considerable challenge and will undoubtedly increase the heterogeneity of the EU (World Bank 2000). We confirm this general statement with our empirical research most of the Eastern European applicant countries lag far behind the EU member states in terms of economic development currently and, therefore, the level of heterogeneity and disparities will increase significantly after enlargement in the union.

Conclusion: Further rural development research tasks

While overall economic trends in CEE countries show a reverse of the decline of the early 1990s and current growth rates of around five percent of GDP, the national figures conceal major regional disparities, for example in terms of unemployment and economic growth. We reveal these regional disparities using various indicators and discuss their consequences on the national as well as the union-wide economic growth in our research. On the other hand we found out, although, striking inequalities between rural and urban areas and within and between different regions characterise almost all applicant countries among the many challenges facing CEE countries today, regional development is perhaps the least researched and least understood. The main tasks we identified for further rural development research are following:

1. Continue and strengthen current research with greater emphasis on integration with *national* experts in policy design and implementation, intensify communication and education activities between current member and accessing countries.
2. Focus effort on providing decision makers with scientifically and sustainably sound decision making tools and policy options to ensure sustainable management of land resources and agriculture.
3. Increase efforts to develop means to support the transfer of knowledge to the Eastern European transition countries to ensure better management of national natural resources.
4. Including expansion of interactions with international organisations and NGOs active in promoting effective and low-cost technology options (forums/conferences/personal interactions).

References

Espoti, R. Sotte, F., 1999. Territorial heterogeneity and institutional structures in shaping rural development policies in Europe, Proceedings of the 9th EAAE Congress: European Agriculture Facing the 21-st Century in a Global Context, Warsaw 24 – 28 August 1999.

European Commission, 1997. Situation and outlook – Rural developments. CAP Working Documents. Brussels.

European Commission, 1999. Sixth Periodic Report on the Social and Economic Situation and Development of the Regions of the EU, Brussels.

Kancs, A., 2000. Modelling Rural Economy: an Inter-regional General Equilibrium Approach, Proceedings of the 24th IAAE Congress: Tomorrow's Agriculture: Incentives, Institutions, Infrastructure and Innovations, Berlin, Germany 13 – 18 August 2000.

Nuppenau, E., A., Thiele, H., 1997. The Political Economy of Rural Restructuring from a regional Perspective: a Neural Network Approach, Proceedings of the 48th EAAE Seminar: Rural Restructuring within Developed Economies, Dijon, 20 – 21 March 1997.

OECD, 1996. Better Policies for Rural Development. Paris, Organisation for Economic Co-operation and Development (OECD).

Swinnen, J.F.M., Dries, E., Mathijs, E., 2000. Critical Constraints to Rural Development in Central and Eastern Europe. Proceedings of the 3rd World Bank – FAO EU Accession Workshop in the Rural Sector: The Challenge of Rural Development in the EU Accession Process. Sofia, Bulgaria, 17-20 June 2000.

World Bank, 2000. Rural Development Strategy: Europe and Central Asia Region. Washington.

Rural Enterprise Planning –
An Applied Approach To Address New Challenges

Balazs, K., Podmaniczky, L.

Department of Environmental Economics, Institute of Environmental Management, Szent Istvan University, Gödöllö., H- 2103, Hungary.

The role of and the approach to the countryside and the environment has undergone dramatic changes in recent decades as relations between sustainable, multi-functional agriculture and preserving natural resources were realised. The Institute of Environmental Management was launched to meet this challenge through training graduate agricultural engineers specialised on environmental and landscape management. These engineers shall have the necessary ecological, agricultural, technical, legal, economic, social and cultural knowledge that enable them to fulfil and co-ordinate various tasks of landscape development, rural development and environmental management.

In order to meet these requirements the curriculum contains basic and advanced applied subjects on natural systems, agricultural production and management, landscape planning in relation to the environment.

Probably the most integrated of all applied subjects is Rural enterprise planning. The objective is for students to acquire relevant skills through such *computerised planning methodology*, developed by researchers of the Department of Environmental Economics that enables them to prepare sound business plans with respect to agro-environmental policy priorities. Such plans may help farmers to exercise effective management controls over the financial aspects of their business through proper planning and record keeping based on calculations of up-to-date information as well as encouraging the introduction and use of farming practices, landscape and landuse planning patterns compatible with the increasing demands of protection of the environment and natural resources (multifunctional agriculture).

Developing problem solving and computer application skills of students is considered very important. The task of students is to prepare a five-year management plan of an existing farm and give a proposal of how the production structure should be shifted towards a "value-conserving" farming system.

The transmission of such knowledge to future rural managers is of paramount importance as there exists a large number of small family-managed holdings where business and production skills are at an elementary level at this stage of the privatisation process and development of rationalised farm businesses within Hungary.

Moreover, application for agricultural state support and planned subsidy priorities of agro-environmental programs will be linked to the condition of having a clear business plan. These facts represent a growing demand for training people for professional rural extension services and an increased need for experts with skills of sound farm business planning.

References:

Ángyán J. - Kiss J. - Menyhért Z. - Szalai T. - Podmaniczky L. - Ónodi G. - Tirczka I. - Kupi K. - Jeney Zs. (1995): Some aspects of sustainable agricultural landscape- and land use in Hungary, Bulletin of the University of Agricultural Sciences, Gödöllő, 75th Anniversary Edition, 1995-1996., Vol.I., 37-50.p.

Ángyán J. - Stefanovits P. - Kiss J. - Ónodi G. - Podmaniczky L. - Győri-Nagy S. - Szala T. - Kulcsár L. - Bakonyi G. - Mézes M. - Keszthelyi K. (1995): New university level educational system for sustainable agriculture and rural development at the Gödöllő University of Agricultural Sciences, Proceedings of Second European Conference on Higher Education in Agriculture, Gödöllő Agricultural University Press, Gödöllő, 123-128.p.

Ángyán J. - van Haarlem R. - Kiss J. - Podmaniczky L. (edit) (1995): Managing Change in the Food-Chain and Environment. Gödöllő Agricultural University Press, Gödöllő, 441 p.

Podmaniczky L. - Ángyán J. - Illés B. Cs. - Straub T. (1997): Farming in protected landscape (economic analysis of the possibilities for sustainable agriculture), IUCN World Conservation Union, Gland (Switzerland), 104 p.

Ángyán J. – Podmaniczky L. (1999): Földhasználat és fenntarthatóság a mezőgazdaságban (Landuse and sustainability in agriculture), Lélegzet, Budapest, IX. évf. 4. sz., 7-8. p.

Philosophic issues behind the curriculum in a new international BSc course in Rural Development and Innovation at Larenstein International College in the Netherlands.

Dunning, T., Hesselink, E.

Larenstein International Agricultural College, 7400 AA Deventer, The Netherlands

The design of the curriculum of the course on "Rural Development and Innovation" is based on the following main principles:

1. Graduates will become **process coordinators** and not just **advisers** for rural entrepreneurs (like farmers) in the field of "rural innovation" (Obs. the term "rural innovation" is also referred to as "rural regeneration" or "rural renewal" or "rural diversification").

2. During the course students "learn to learn". Student-centered learning is applied as radical as possible. For example students determine for about 50 % the curriculum they want to do. Also, all subjects except basics like Information Technology and Language are learned through "project-education".

3. Integration of "learning outside the school" with "learning within the school". Students will spend about 50 % of the curriculum time outside the school, in all kind of supervised placements at organisations involved in "rural innovation".

4. International orientation. It is believed that exposure of the students to international experiences will not only benefit their knowledge and skills with regard to Rural Development/Innovation but also enhance their capacity for self-reflection, a skill and attitude indispensible for their future career.

5. Creativity. A very important asset (but difficult to "learn") for the "rural innovator".

Agricultural Education in transition – benefits and experiences from Case Studies

Contributed papers

Higher Agricultural Education and Research in Latvia - Results of Reforms and Future Challenges

Busmanis, P.

Latvia University of Agriculture, Jelgava, LV-3001, Latvia

Transition reforms in Latvia after the restoration of independence in 1990 are going through basic changes in the agricultural sector, what expriences essential decreases in the production level. At the same time, agriculture as a part of rural development has good environmental and social preconditions for economic development in sustainable way, what should be based on wide education and research.

In the last decades wide reforms have taken place in Latvian higher education and science. These changes have influenced the content, structure and administration of academic institutions, the number of personnel, sources of funding etc. With the new economic conditions the role and function of education and science in the community has changed, as have its goals and tasks. Research and higher education has to be oriented to the objectives of sustainable rural and national development: effective economics, social welfare and saved environment on parity principles. The Latvia University of Agriculture (LLU) as a leading agricultural institution in Latvia plays a decisive role in solving these tasks.

Higher agricultural education in Latvia first was commenced in 1863 at the Riga Polytechnical Institute, which was the first higher education institute in Russian Empire of such kind. After the formation of an independent Republic of Latvia, the University of Latvia was established in 1919, including also the Faculty of Agriculture. The Latvia University of Agriculture as an independent higher education institution was established in 1939 on the basis of the Faculty of Agriculture and Forestry of the Latvia University. After World War II higher education and science in Latvia developed under control of the Soviet Union's governmental and administrative bodies.

The higher education and research **policy,** implemented in Latvia after the regaining of independence has been determined to a great extent by the conditions of the transition period. The radical changes in politics, economy and culture have created a totally new environment for the development of higher education and research. As a result, universities have become more autonomous administrative structures.

A new **legislation** of education and science was developed. The first Education Law (19.06.1991) with 10 paragraphs pertaining to higher education allowed commencing reforms. The next Law "On Higher Education Establishments" (1995) determines essential further changes in all spheres of higher education. The Law "On Scientific Activity" (1992) with amendments (1998 and 1998) regulates administrative, financial and institutional conditions in the area of research. The National Concepts of Higher Education and its Development (1998) and National Concept of Research Development (1998) was formulated and accepted. The strategy is worked out for the period up to the year 2010 and it takes into account the growing role of education and research in society and its impact on economy. The Rural Development Program of Latvia (1998) was elaborated to create preconditions for integrated, multiform and sustainable rural development based on diverse education and research. Agriculture and forestry as part of rural infrastructure might play a growing role on the management of natural resources.

Organisation and administration of agricultural sciences and higher education is similar to that commonly established in Latvia system, with its own specifics. Traditionally the Ministry of Agriculture is responsible for agricultural research and education. Under the authority of the Ministry are the Latvia University of Agriculture, two research institutes, state experimental and breeding stations, agricultural colleges and schools, Latvian Agricultural Advisory Service. The Latvian Academy of Agriculture and Forestry Sciences was founded in 1992 as an association of the most prominent scientists in the related fields of these sciences.

Special emphasis was paid to the development of the unity of research and education at the University. The process of formal integration between LLU and four state research institutes was carried out at the beginning of 1998 and today they are Research Centres of the relevant faculties. A complete functional integration is still the matter of time and subjected to financial and managerial stimuli.

Reforms in higher agricultural education and research during the last decade after the renovation of Latvia's independence can be evaluated as successful, especially if considering the very limited financial and academic resources from the transforming economy. Still, future national and international challenges create the necessity for further development of higher agricultural education by:

- provision of high quality agricultural education and promotion of retraining from conventional agricultural professions to alternative professions related to diverse rural entrepreneurship and natural resources management;
- diversifying of study programmes, improving curricula and syllabus, introducing new didactics and methods of studies;
- paying more attention to continuing education and extension activities;
- further developing international dimensions, providing full-bodied bilateral mobility of students and professors.

Future models of agricultural higher education should go ahead of political, economical and social changes.

References

Busmanis P.1994. Latvian Rural Development and Agricultural Education. In: Managing Change in the Food-chain and Environment: the Role of Higher Agricultural Education. Abstracts of Second European Scientific Conference on Higher Education in Agriculture. Gödöllö University of Agricultural Sciences (Hungary), pp.22 - 23.

Busmanis P.1995. The Way Towards Market Economy and the Reforms in Higher Agricultural Education in Latvia. In: Strengthening Teaching and Developing Training Materials in Socio - Economic Subjects in Higher Education Institutions in Central and Eastern European Countries. Report on FAO/WAU workshop, Volume II. Rome: FAO, pp. 23 - 38.

Busmanis P.(1999), Environmental situation in Agriculture. Country Report of Latvia. In: Present Environmental Situation in Agriculture. Country Reports for the First Workshop of Central and Eastern European Sustainable Agriculture Network. FAO, Humbolt University of Berlin, Godollo, Hungary, p.115-152.

From agricultural to agribusiness and rural management education: a case study from Poland's transitional economy and university environment

Figiel,S., Pilarski, S., Warzocha,Z.

University of Warmia and Mazury in Olsztyn, International Center for Business and Public Management, 10-720 Olsztyn, Poland

Free market oriented economic reforms, which have been implemented in Poland since 1990, among many other fundamental changes induced a tremendous demand for higher education. In the 90-ies the number of students in Poland rose more than 3.5 times (from 0.4 to over 1.4 million). This rather unusual increase in the total number of students in Poland was accommodated not only by larger enrolment at the public schools but more importantly thanks to a rapid development of private colleges. However, the demand for various areas of studies and forms of education appeared to be very different as compared to the one in the centrally planned economy system. The process of market transformation revealed simply different structure of demand for education than the previously existing educational capacity in the country.

Most of the new demand for higher education occurred in the business management area, as such skills became highly desirable in the radically changing economic environment. Effects of these changes were much more profound for the universities heavily involved in higher agricultural education than in case of many other schools. They experienced considerable decline in number of candidates interested in studying agriculture sciences as well as lack or very little market recognition for the skills of their graduates.

The purpose of the paper is present main features of the transformation process, which has taken place in the last seven years in the area agricultural economics education at the University of Agriculture and Technology in Olsztyn. The University is located in the north-east part of Poland where problems of restructuring the former state agricultural sector and rural unemployment, exceeding in some areas 20% of the total workforce, are especially severe. Such situation constitutes a serious challenge not only for the policy makers but the school itself facing radical changes in the job market.

The educational experience discussed in the paper is mostly based on the development of two executive programs, namely non-degree one year Post Diploma Studies in Marketing and Management (PDS) and two year Executive Master in Rural Industries Management (EMRIM). The process of implementing them and the impact they had on the students are described in greater detail by looking at some aspect of selected professional career stories.

The basic analytical tool used in the research was comparative analysis applied to the changes in the university structure observed over time using available university statistics with special focus on agricultural economics education and development of the new program curricula. To gather relevant data and information about the graduates from the analysed programs a survey technique was used. Also personal interview was used to describe the impact of the programs on some professional careers of the graduates.

At the beginning of the 90-ies Olsztyn University of Agriculture and Technology (OUAT) had 8 faculties where about 12 thousand students studied agricultural sciences and technologies related to agricultural and food sectors. At same time the demand for such specialists started dramatically to decline, as other skills became much more valuable in the job market increasingly driven by rapidly emerging private businesses. This refers mainly to business management skills, which in general were hardly taught at the universities under centrally planned economy system.

Under such circumstances the OUAT decided to restructure and offer much broader variety of majors and options to study. One of the areas recognised as most important was management education built mainly upon existing teaching capacity in agricultural economics. The development was two directional, namely establishing regular major in management for full time students, and creating post diploma studies programs offered to university graduates interested in attaining professional management skills. The latter educational offer was meant especially for all those who already had managerial positions in various companies and organisations but had never received formal economic or management education. PDS and EMRIM are good examples of such retraining programs.

The PDS curriculum consists of 220 hours of instructions and includes such topics as principle of economics, business application of quantitative methods, managerial economics, financial management, operations and strategic management, business negotiations, market analysis, marketing research, international marketing and marketing management. Since the EMRIM program is a master degree program its curriculum is more extensive (540 contact hours) and covers the following topics: micro and macroeconomics theory, managerial accounting, quantitative methods, managerial economics, business law, business communication, consumer economics, finance, price and market analysis, operations and strategic management, human resource management, food marketing management, environmental and natural resource economics, international economics and trade, community and regional development, business English and master thesis seminars. It should be mentioned that the program curricula as well as teaching capacity and materials were developed with a great deal of foreign assistance. Both programs focus on teaching practical managerial skills and the use of computers. Teaching materials and cases source examples mainly from agricultural and food sectors and rural economy in general.

Since 1993, 325 persons from the PDS program have graduated. The EMRIM program was launched in 1995 and till now there are 75 graduates of it. In April 2000, all the graduates were asked to fill an anonymous questionnaire meant to highlight the role of the programs in their professional careers. 101 of the PDS graduates and 30 of the EMRIM responded to the survey. Whereas majority of them work in rural areas only very few in the agricultural sector. According to their opinions 80% of the respondents became very well paid managers in various companies. 8% of them established their own businesses. Some of the main advantages gained from the programs as they pointed out are the following: enhancement of professional skills (86.6%), better communication with others (60%), ability to solve company problems (43.3%). From various areas of managerial knowledge and skills covered in the program they found marketing as the most valuable (about 60% of the respondents).

Educational experience presented in paper proves that agricultural university can successfully meet new educational demands generated by transformation of the economic system by developing new curricula and implementing relevant programs. A very important aspect of this approach is to target not only regular daily students but also those with university degrees who require retraining because of the changing needs of the agricultural sector and the whole rural economy. This can be achieved much easier if the institution is dedicated to change. In fact in the last decade the OUAT went through a fundamental restructuring process, which after a merger with the Olsztyn Pedagogical College eventually led to the establishment of a full fledged university in 1999, namely the University of Warmia and Mazury in Olsztyn (UWM). The University now has 12 faculties and 24.5 thousand students.

Strengthening the co-operation between higher education and rural economy continuing education and integrated extension – curriculum development in two Albanian Universities

Bicoku, Y.,[1] Androulidakis, S.,[2] Phelan, J.[3]

Institutional Support Unit, International Fertiliser Development Center, (IFDC) Albania.
[1] Deputy Chief of Party AAATA/IFDC-Albania, [2] Professor, Technological Educational Institute of Thessaloniki, Greece, [3] Professor, University College Dublin, Ireland.

This paper deals with the joint efforts of seven partner institutions (three of EU and four of Albania) in an attempt to meet the objective "Development of curricula and courses on Agricultural Extension Training in higher education for achieving the improvement of responsiveness of undergraduate education to changing environments and equip the graduates of such schools with competencies that will be practical for their placements in Public or Private extension services.'

The driving forces for this project were certain priorities set for Albania:

- Strengthening the cooperation between higher education and industry/economy;
- Development of curricula and courses in order to achieve improvement of responsiveness of undergraduate education to changing environments of the economy, and
- Introduction of adult continuing education.

The target groups of the project are (a) the Agricultural University of Tirana and the F.Noli University of Korca, the staff of which needed to be trained in offering extension, (b) the Ministry of Agriculture and Food which needs to train on extension methodology more than 700 of its staff working as extensionists, and (c) the Albanian Fertiliser Ag-input Dealers Association (AFADA) which has undertaken to train their dealers as private extensionists having for this reason developed a Private Extension Service agency sponsored by the International Fertiliser Development Centre (IFDC).

According to the above, the wider objective set was to ensure that the two Albanian Universities with agricultural faculties are in position to offer agricultural extension education which is targeted towards the changing needs of the rural economic environment.

In meeting this, the project's objectives included:

- To produce a complete set of course curricula, subject matter and teaching/training material for both the undergraduate and the continuing education courses;
- To facilitate the setting up of an extension laboratory in each of the two universities and equip them with the necessary equipment;
- To extent this laboratory as an information on agricultural issues center;
- To produce a video on extension that can provide information on extension issues for those interested as well as to become as an extension teaching tool in the future.

The activities realized in fulfilling the above, and the contribution of each institution included:

- Two seminars in Greece provided by the staff of the two Greek partner Institutions (Technological Educational Institute of Thessaloniki, and the Development Agency of Karditsa ANKA);
- Two seminars in Ireland provided by the Partner Institution (University College Dublin);
- Participation at the Extension Conference for Former Eastern European Countries held in Eger, Hungary;
- Two working seminars in Albania;
- Two evaluation meetings in Albania (one each year) giving emphasis in making the public aware of this project;
- Several visits to Albania by the project coordinator in facilitating and coordinating all activities.

Difficulties faced include; (a) the instability of the situation in Albania during, (b) the lack of smooth continuation of the project due to staff changes, and (c) the shortage of money provided by the European Training Foundation Centre since the whole project was a TEMPUS University Management Project and it is financed by the Centre.

Recommendations will be provided at the end of the Project (March 2001) and include: (a) to the Universities the responsibility to make good use of the project's outcomes, and (b) to the Ministry of Agriculture and Food to take advantage of the continuing education curricula in employees and in return they will apply life long learning for their clients the farmers order to train those of the Ministry's staff dealing with the advisory service.

Thus, these institutions will also help for the retraining of public and private extension of Albania. Next step to be followed is to develop continues training courses for enterprises and other organisations.

The Institute of Economic Studies and Transformation of Economic Education at Slovak Agricultural University in Nitra

Bartova, L.[1], Tauer, L.[2], Bandlerova, A.[1].

[1] College of Economics and Management, Slovak Agricultural University, Nitra, Slovakia
[2] Dept of Agricultural, Resource, & Managerial Economics, Cornell University, Ithaca, USA

The Institute of Economic Studies (IES) was established at Slovak Agricultural University, Nitra (SAU) in 1993 as a joint project of Cornell University, USA (CU) and SAU funded by Andrew Mellon Foundation. Faculty of the Department of Agricultural Resource and Managerial Economics at Cornell (ARME) taught a full set of course offerings in a one-year program at the Masters' level. Grant lasting for 2 academic years enabled training students to be better able to understand economic development. Selected students came to CU to complete their MS study.

The purpose of second IES project from 1996 to 2000, funded by Andrew Mellon Foundation was to assist in the professional development of the economic and social science faculty SAU. This was accomplished by SAU faculty visiting Cornell University for extended periods during the academic year, and by holding a series of workshops in Nitra on economic research topics. SAU faculty visited Cornell University for study leaves of 8 weeks to a full semester. During those visits SAU faculty were appointed as Cornell visiting scholars, and were given access to University facilities. They attended regular academic classes and seminars. Many worked on new courses, while others worked on research reports.

A total of 8 workshops were held at the Slovak Agricultural University on various topics. A team of two Cornell faculty members taught each workshop, which were one week in duration. Besides SAU faculty and graduate students, these workshops were opened up to faculty in other Central and Eastern European countries. Opening up these workshops to the region not only provided a professional development opportunity to all regional faculty members, but also provided opportunities to those visiting the faculty to meet and work with the faculty of SAU.

The major accomplishment of the joint project was the enhancement of the teaching and research processes at SAU. Participants, either in the workshops or by visiting Cornell, observed economic concepts being used to analyse economic problems, both in transition economies and/or full market economies. Slovak faculty revised and started new courses using the materials and research results obtained. Faculty members have a better appreciation of the skills that students need in order to succeed in a world economy. These students will become increasing capable of good decision making both in businesses and in politics. In the workshops participants learned the process of research problem identification, selecting appropriate analysis techniques and objective reporting, (i.e., how to do research). Participants of seminars from the various universities and research institutes from different countries made profession connections, and are able to work on common problems facing their countries.

We will see increased research activities from Slovak Universities in the years ahead. The limitation now is the lack of research funding to these Universities from the government, and the heavy teaching responsibilities of the faculty. Since external grants and contracts will become a more important source of funds for Slovakian Universities, it was useful to observe and participate in the procedures used to obtain and administer externally funded projects.

Although additional work focused on teaching would be useful, future activities should focus on enhancing the research capabilities.

Refocusing higher agricultural education to address rural development in South Africa

Linington M.J., Ferriera F.M.

VUDEC, Departments of Agricultural Sciences and Community Resource Technology, Vista University, Pretoria 0001, South Africa.

Vista University is unique in South Africa in that it is a multi-campus university. It historically, has seven campuses offering contact tuition and one campus offering distance education tuition. At present we have 20,000 students, 50% enrolled via DE and 50% enrolled for contact tuition. The department of Community Resource Technology and Agricultural Science within the Science Faculty of Vista are unique in that they are one of the few University Departments in Africa to offer their courses via distance education. In fact both these departments offer courses in DE mode only.

The education scenario in South Africa has changed drastically since 1994. Curriculum 2005, an outcomes-based form of education (OBE), has been implemented at schools. Tertiary institutions are being required to resubmit their qualifications in OBE mode to the South African Qualifications Authority (SAQA).

Historically, the Department of Agricultural Science at Vista University offered only teacher diploma courses for in-service retraining of teachers. The changing educational scenario in tertiary education in South Africa seemed like an ideal opportunity to reposition ourselves in terms of our offering and our content. Starting this year we have been involved in the implementation of three new degree programs, which are unique in both content and design. The goals and objectives were:

1. To introduce high quality qualifications which are applied and implemented within resource poor areas
2. To encourage the student to be innovative in implementing knowledge to the benefit of both the community and the environment.
3. To focus our research resources on urban agriculture.

The three degrees introduced were: B.Sc. Ed. (Agric.) degree programme, a B. Human Ecology (Community Agriculture) degree and a short course in Food Gardens was introduced to a B. Human Ecology nutrition degree to form a B. Human Ecology (Community Nutrition) programme. Finally, diplomas were designed as exit levels of the degrees.

The uniqueness of our programmes cannot be seen in the name but in the content. The scientific content in terms of maths, chemistry and physics was cut down to the level of need to know and not nice to know. All courses teach scientific concepts and principles at tertiary level while requiring the student to apply it at community level. The importance of the environment and ecology in agriculture is integrated at all levels of the programme. How we overcome the limitations of correspondence learning and teach these concepts on a distance base will be discussed.

Finally, we will look at the world-wide trend in extension to combine production with consumption, i.e. agriculture with home economics.

Farm Animal Behaviour and Zoopsychology of the Domesticated Animals – A New Course for Russian Agrarian Universities

Georgievsky,V.I., Ivanov,A.A., Sheveljev,N.S., Afanasjev,G.D.

Timiryazev Agricultural Academy in Moscow, Russia

The main idea that unites disciplines in the traditional animal science curriculum is farm animal performance. High dairy productivity, high meat, egg, honey etc. performance of animals is the condition of successful farming. It makes animal husbandry profitable. The curricula of animal science universities and colleges are built in accordance with this fundamental. The traditional curriculum is composed of four blocks of disciplines:
- -General (philosophy, foreign language, mathematics, chemistry etc.),
- -Biological (anatomy, physiology, genetics etc.),
- -Technological (animal feeding, herd management, sire tests etc.),

-Economical (animal husbandry economy, bookkeeping, marketing etc.).

The total number of disciplines the animal science student is offered during his five-year stay in a Russian university exceeds 60. Nevertheless many active students and professors are not satisfied with modern animal science curricula.

The economic situation in Russia has changed dramatically during the last 5-7 years. The year our students reproach us for the *inhuman* treatment of farm animals. Most of our colleges market of agricultural products is overloaded. Russia's agriculture is facing new problems: the problem of quality, the problem of ecology and geography of agricultural production. Moreover, the consumers of animal products put questions to producers about the welfare of farm animals and animal rights. Our students don't want any more to work on farms where animals are badly treated. Among animal product consumers there is a strong opposition against animal products received from animals that experienced physical sufferings. On the other hand, farmers began to appreciate the benefits of animal welfare.

Each school accept this reproach. But the problem is that very often we do not know what is human and what is inhuman in our relations with farm animals. We understand that to feed a cow with biscuits and coffee is inhuman. It is more human to use straw and hay in feeding a cow. We realise that for most of the people a visit to a slaughterhouse is a serious trial of their mentality and psyche. On the other hand, many people are absolutely indifferent to artificial insemination of females. Though this procedure has much in common with woman violation.

So, in our estimation of technological methods accepted in animal husbandry we must use some criteria of what is human and what is not. Incidentally, this is not only a moral aspect of animal science. Animal sufferings are the result of stress-factors impact. It is well known that stress through activation of the sympath-adrenal complex brings about widespread neurological and endocrine responses that, over a period of time, cause changes in the functioning of many body organs, often leading to disease, performance reduction. Finally, inhuman treatment of farm animals reduces the economical efficiency of animal husbandry.

As a criterion of human treatment of farm animals we us animal welfare which we define as a condition of full satisfaction of the main physiological demands of farm animals (nutritional, sexual, social, maternal, environmental). In order to determine whether an animal is in a state of physiological comfort or not, a farmer (pet owner) has to be aware of the main behavioural stereotypes. The knowledge of zoopsychology contributes to the better understanding of the animal behaviour. Psychology in its tern presupposes the understanding of the motivational mechanisms in the Central Nervous System of animals. So, the introduction of new disciplines - ethology and zoopsychology of the domesticated animals into existing curricula has become inevitable. Basically, the goal of these disciplines is to provide a student of animal science with the knowledge, practical skills of environmental management and personal qualities to reduce farm animal sufferings as much as possible and finally to rise animal husbandry efficiency. The programme includes several parts, which can be offered either separately as independent disciplines or within existing courses (zoology, morphology, physiology and biochemistry). But in any case the programme is designed for the undergraduate students who have coped with the first two-year curriculum. Students of the third year would need about 70 academic hours of classroom studies to qualify the final test. The whole course may consists of several modules like this:

1. Ethology: Contents, history, and methods of investigation. Inborn forms of behaviour. Classification of instincts. Conditional and unconditional reflexes. Animal learning (active and passive forms). I.P.Pavlov's concepts. Biological and social aspects of animal behaviour. The role of genetic and environmental factors. Complicated forms of farm animal behaviour. Behavioural peculiarities of different species of farm animals. Individual behavioural peculiarities of farm animals. Pavlov's types of high nervous activity.
2. Central morpho-functional correlates of a behavioural act: Reception, associations, central inhibition, the principles of co-ordination. The theory of dominant. Memory (types and possible mechanism). Theories of memory. Sleep (rapid and non-rapid), dreaming. Theories of sleep. Morpho-functional analysis of animal behavioural stereotypes. Concept of functional systems of P.K.Anokhin.
3. Consciousness (psyche). Levels of consciousness. Comparative evaluation of the integrative function of brain. Psychic and somatic unity of animal organism.
4. Moral and ethic aspects of man and animals interrelations: Farm animal welfare (metabolic, behavioural, social comfort). Animal rights and juridical defence of farm animals.

The suggested modules are independent enough. So these modules can be worked out and presented to students by several experts: zoologist, physiologist, psychologist, and lawyer (philosopher).

Facing change in the curriculum and its delivery

Contributed papers

Farm Management: the Death of a Discipline?

Warren, M.

Seale-Hayne, Faculty of Agriculture, Food and Land Use, University of Plymouth, Newton Abbot, Devon. TQ12 6NQ. United Kingdom.

A casual glance at the Agricultural Higher Education sector in the United Kingdom shows a dwindling number of academic staff specifically identified with farm management, as well as fewer courses and modules in the subject. Faculties which in the past were proud to flaunt their farm management expertise – and which indeed often relied on farm management staff for industrial liaison and public relations – now seem barely able to tolerate it. Are we seeing the death of a discipline? If so, why? – and does it matter?

This paper firstly seeks to establish whether the issue is more apparent than real, and then investigates some of the possible causes of the observed signals. One of the influences may well be the general move towards integrated rural development, the theme of this conference: as we think more in terms of rural business rather than agricultural business, the rationale for treating farm management as distinct from general business management comes increasingly into question. Given that more and more non-agricultural faculties (in, for instance, business studies, geography and human sciences) are turning their attention to agriculture, we may be witnessing the decline of departments rather than of a discipline. The lack of strong leadership from a professional body (in contrast, for instance, to the Royal Institution of Chartered Surveyors) may be a factor.

In its conclusions, the paper will argue that there is still a need in the curriculum for specific farm management approaches, built on a sound understanding of general business principles and techniques. Without a strong research base, though, farm management cannot survive as a separate discipline in today's Higher Education system. Key to this are the identification of a research agenda, real collaboration within and between disciplines, and effective dialogue with professional managers in the industry. The paper will not attempt, however, to present a definitive answer, but will aim to create a basis for discussion and, perhaps, action. In particular, it is hoped that the discussion will draw on the experiences of delegates from other parts of Europe.

Developmental Tendencies of Agricultural Education in Poland

Wieczorek, T.

Warsaw Agricultural University, Department of Human Sciences, Poland

1. Since 1989, when the transformation process was initiated in Poland, it has involved many fields of social and economic life, including agricultural education. Agricultural schools have faced the task of preparing graduates for work in agriculture and services for the rural sector under conditions of market economy. Education must get rid of all relics of the centralized education system which limits initiative, individualism, empowerment and identity of the school, teachers and students.

2. The shift from command economy to market economy has put the rural sector and agriculture in Poland in completely new situation. Uncertainty of prices, drop of demand for food, liberalization of foreign trade, abolishment of subsidies to means of production and services, and increase of their prices, relatively expensive credit – these are just some of the macroeconomic conditions to which agriculture had to adapt.

3. The social and economic situation of the rural sector and agriculture, changing in line with the development of the foundations of market economy, calls for changes in agricultural education. The modernization process in agriculture and its environment will largely depend on the level of qualifications of agricultural producers and the employees of the food sector in general. In the economic situation of the country, the farmer is required not only to have an agriculture expertise, but also the ability to think in economic terms and cope with changing circumstances. Thus agricultural education now faces urgent need to reorient its existing nature and scope of economic training of graduates.

4. The changes in the agricultural education system must involve technological and technical modernization, public awareness building and development of ethic and moral attitudes. In addition, the transformations taking place in agriculture and its environment must be taken into account. Market economy imposes ownership changes, makes it necessary to increase production and develop services. Under such circumstances schools are expected to demonstrate full responsibility for the vocational training and comprehensive development of their students.

5. The school should also have a supportive role vis a vis the region. This requires joint decision-making in selecting the educational targets and contents by local governments and agricultural organizations. The school has an increasingly important role in professional development of farmers and dissemination of progress in agriculture. There is also an acute need to incorporate agricultural schools in cohesive agricultural expertise system which is formed by research institutes on the one hand, and agricultural counselling centers, farms and firms supporting agriculture on the other hand.

6. The key factors in modernization of the Polish rural sector and agriculture will be education and development of agricultural processing. Poorly educated rural population making a living out of agriculture is not able to properly manage farms or find employment outside agriculture. On the other hand, processing means an injection of new know-how into agriculture, more advanced culture, modern standards.

The Polish rural areas suffer from a shortage of educated people with farming skills. This is confirmed of the following data. In urban areas, 17% of the self-employed have a higher educational background, 40% have secondary, and just 23% primary or incomplete primary education. Among the farmer population, 0.5% have higher education, 13% - secondary, 51%

- primary, and 7% - incomplete primary education. Thus almost 60% of farmers have no professional qualifications. In a town they would be unskilled laborer, while they should be employees, businessmen, marketing specialists and accountants. No wonder then that they cannot cope with the requirements of modern farming.

7. The needs of a free market economy require that the training of specialists be abandoned in favour of a broader agricultural preparation. Introduction of such broad profiles of teaching into the universities requires a new and complex approach to: qualifications of graduates; study plans and study programmes; books, textbooks and supplementary materials for teaching professional subjects; and conditions of realization of the teaching process with particular consideration for the practical preparation of students.

8. Agricultural universities must fundamentally reorient their positions as regards the type and range of economic training of students. The situation has changed profoundly. An appropriate preparation in economics of graduates of agricultural universities is required due to the pro-market orientation in farming. Without knowledge of economics a farmer is helpless when the rules of a consumers' market determine the size, structure and techniques of production and consequently set production costs. Moreover, in a free market the economic condition of a farm is affected by its ability to compete with local and foreign agricultural producers on both the national and foreign markets.

9. Private farms will create a growing and varied demand in the labour market for graduatesof agricultural universities. Potential heirs of farm owners and future employees of agricultural service industry will study agriculture. The following professions will also require graduates of agricultural universities: counselling for agriculture, animal husbandry, economics and the organization of such services; ecological watchdogs and their agent; and specialized farming and animal husbandry services in unions of producers and breeders in agricultural self-government.

Training for the above-mentioned professions is fundamentally different from that for specialized workers in large state-owned farms. The new graduates must have a broad training in natural sciences, technology and economics. They should also be well trained in ecology, know the principles of free market economy, be able to use new techniques and be able to use sources of information about the latest scientific discoveries.

10. The current model of education under the M.Sc. programme has been undermined by the situation in the labour market. Economic and social needs require educational programme which can fulfill various functions by offering different degrees and levels of qualification. Agricultural studies should be based on the principle of multi-level system. Besides the traditional masters, doctorate, and engineering degrees agricultural universities should also offer short training courses intended for people who want to: switch professions or specialisations, increase or update professional knowledge, and obtain new information. Organizational flexibility must be matched by flexibility in programmes, thus making it possible to tailor subjects not only to local needs but also to student possibilities and interests. Modern agricultural university should limit the number of compulsory subjects for the sake of electives or only partly limited choices. This is the way to promote individual study, which is well served by module studies based on credits.

Agribusiness Management Development Program: Stimulating Rural Development by Facilitating the Entrepreneurial Spirit in Rural Saskatchewan

Cooney, A.

Department of Agricultural Economics, College of Agriculture, University of Saskatchewan, Saskatoon, Saskatchewan, Canada

Production, processing, transportation and information technologies are revolutionising the agriculture and agribusiness industries around the world. Some major trends in agribusiness can be observed 1) as domestic and foreign consumers receive information on the production of their food, they are becoming more discriminating and demanding on the issues of food safety and nutrition, and on methods of production and processing; 2) governments, consumer groups, supply chains and individual firms are working to improve the integrity of the food supply chain; 3) producers and processors are working to increase market share in domestic and global markets; and 4) there are international changes in policy and trade which are changing the nature of agribusiness. The continued industrialisation of agriculture is a fancy way of saying that we are moving from a picket fence mentality to a boardroom mentality. This move requires individuals in the industry to have a whole new set of entrepreneurial skills and business skills. The dynamic nature of this changing environment requires that agribusiness professionals receive ongoing training in both production and business.

The province of Saskatchewan, Canada, has a large primary agriculture sector with little value added processing. Farm production has shifted from the subsistence family farm which received extra income from the sale of surplus commodities to the family farm which relies primarily on the sale of commodities. It is now shifting to the entrepreneurial or commercially oriented farm enterprise which is more aware of its role in a value chain. The shift to an entrepreneurial focus in commodity production or through the diversification of products or farm service businesses is an arduous process. Government support for this paradigm shift is lacking or inappropriate for the agribusiness entrepreneur. The University of Saskatchewan and The Agriculture Institute of Management in Saskatchewan (which delivered introductory level business training to farmers), saw that educational support was required to provide progressive farmers and agribusiness professionals with tools and training to update and refresh business skills while focussing on the changing agribusiness environment.

An executive management program for agribusiness professionals was developed by professors in the College of Agriculture, College of Commerce (business school) and the Extension Division who recognised that it is the University's mandate to serve all stakeholders including mature students and those in need of training. The Agribusiness Management Development Program (AMDP) aims at those with lots of experience in an agribusiness with or without an undergraduate degree. It is a leading edge program with a focus to value added, from farm gate to plate, which enables participants to apply modern theories of agribusiness management to their own operations. By its nature, the program integrates agriculture, business and experience and builds on previous agriculture and business education. AMDP adds a human resource management component which is missing in most undergraduate degrees. Marketing is also handled in a unique manner as it provides entrepreneurial assistance rather than encouraging a reliance on marketing boards and organisations. The program provides management skills which enhance rural economic development initiatives.

The program is delivered in four modules which are held on four weekends over a period of three months. Program information is progressive and participants are encouraged to attend all four modules. This schedule minimises the time away from business and family while immersing participants in the course content. The time between modules allows them to think through the concepts learned and to apply them to their enterprises.

The participants are adult learners and for this reason case studies are incorporated into presentations, reading is distributed before a module is held and copies of overhead materials are provided to decrease note taking and to encourage thought and participation.

Core areas taught are: Finance/Accounting, Human Resource Management and Agribusiness Marketing. Special topics to address current issues such as food safety, investment analysis, Saskatchewan Aboriginal People in the 21st Century are also included in the program. The modules are highly interactive and competitive. Participants learn to develop and manage all components of a business plan and they compete in an agribusiness simulation exercise.

The four program modules are:

Module 1: The New Era In Agribusiness Management
- provides an in-depth look into the trends and challenges facing agribusiness today

Module 2: Achieving Goals and Growth Through Effective Agribusiness Planning Strategies
- teaches participants to successfully develop and manage all components of the strategic business plan to maximise opportunities in expanding domestic and global economies

Module 3: Integrated Agribusiness Management
- operationalises all components of the agribusiness plan in a hands-on agribusiness simulation in a competitive environment

Module 4: The Globalisation of Agribusiness
- focuses on issues facing entrants in the international market place

The program aims at a broad range of agribusiness managers (farmers, lenders, consultants, etc.) because it was recognised that with the trend to vertical alliances in a changing global and domestic market place, there was a need for all sectors within the industry to form alliances and to understand all aspects of agribusiness management to make the best possible decisions. Participants have embraced the networking opportunity with other leaders and entrepreneurs. They have found new business contacts, learn from other management perspectives and gain a wealth of experience and ideas from other agribusiness leaders. This is especially important as the trend to larger farms and rural depopulation decreases the opportunity for entrepreneurial spirits to meet and support one another.

The Agribusiness Management Development Program was started in 1999. Participants from 1999 and 2000 have formed an alumni committee to oversee the development of alumni activities and to support the creation of a second level of training for alumni. Program details can be obtained at the website:
http://www.ag.usask.ca/news/conferences/amdp/

# Educational improvement and innovation in life sciences and nature conservation. Key factors for safe food in a sustainable world.

Driesse, J.

Holding Van Hall – Larenstein Agricultural University, Velp, Deventer and Leeuwarden, The Netherlands.

1. Specific tasks for the social and scientific sectors of agriculture, nature-management, food production and animal sciences (the agro-nature-complex):
   - sufficient and safe food for a rapid growing world population
   - substantially more sustainable production-circumstances
   - reintegration of nature-conservation in agricultural practices
   - growing quality demands for food and food-production methods
   - overall conservation and enlarging natural areas, flora and wildlife
   - sufficient financial returns on economic activities in the agro-complex.

2. Elements of the Dutch agro-nature complex
   - operating in a broadening sector: agriculture – horticulture – nature – animal care
   - high production volume, high export-performance
   - high productivity
   - strongly developed quality-systems and quality-awareness
   - strong homogeneous knowledge-system (research – development – education and extension).

3. Reasons for this position:
   - natural circumstances
   - entrepreneurship
   - organisation of the farmers and companies
   - governmental subsidies and support
   - flexible and robust knowledge system.

4. Schematic view of the Dutch agro-knowledge system.

5. The role of the knowledge-system in meeting the social tasks and challenges.

6. Specific position of the Holding Van Hall – Larenstein:
   - universities for professional education in agribusiness, nature, life-sciences. About 5000 students
   - regional knowledge-centers.
   - compact and flexible organisations
   - open and short communication-lines with the professional groups
   - part of the knowledge-system. Access to science-development in cooperation with Wageningen University and Researchcenter
   - fast implementation of developments and innovation in the educational content and consultancy.

7. **The educational system of Holding Van Hall – Larenstein**
   - aimed at the fields of interest and feelings/trends among young people:
   - nature, animals, safe food, life sciences
   - integrated modular system; partly problem-oriented
   - integration of general professional skills
   - integration of sustainability and nature-conservation
   - delivering many possibilities of choice, especially aimed at a development from personal interests and motivation to a professional package of knowledge, behaviour and skills. Such in relation to concrete professional profiles
   - Strong international orientation. Delivering different courses in English language.

8. Further goals and challenges for the Holding Van Hall - Larenstein
   - broadening the cooperation with Wageningen University and Researchcenter. Delivering parts of the Wageningen-curriculum in Leeuwarden
   - substantial increase of the number of students
   - increase of English-spoken courses and number of foreign students
   - developing and improving new curricula: animal management, rural development, marine studies
   - massive investment in distance-learning and integration of Information and Communication Technology in the curricula
   - specific job-policy, aimed at a high "labour-market-efficiency". By means of an agency for finding jobs and employment and deployment of temporary workers increasing the share of graduates that gets a job in our sectors.

9. Conclusions
   - The key-role of Professional Universities in The Netherlands is that they deliver graduates that are extremely properly educated for functions in the interface between production and innovation
   - Professional universities can deliver a substantial contribution to the necessary changes in the agribusiness and nature- and food policy
   - They can do that also as a knowledge-center that delivers market –oriented knowledge and skills in different forms (courses, consultancies, application and development, etc.)

References
   - Businessplan Holding Van Hall – Larenstein; December 1999
   - Power and Quality, policy-document Dutch Government, Department of Agriculture, Nature Management and Fishery, March 1999

Developing the pedagogical competency of a university lecturer

Slavík, M., Miller, I.

Department of Education, Czech University of Agriculture, Prague, Czech Republic.

The paper analyses the functions of university lecturers in agricultural education with reference to their pedagogical competency. It presents the results of individual assessments of personal pedagogical competency of more than 150 academic staff of the Czech University of Agriculture (CUA) at Prague. It also discusses the experience gained with training courses to

According to the classical understanding of university activities the mission of a university lecturer is the pursuit of research, develop pedagogical competency, which have been organised by the Department of Education at CUA for young lecturers in all Faculties of the University. On the basis of the results obtained, as well as what has emerged from other analyses, it raises for discussion some items, and their mutual connection and importance have often been discussed in professional environments. Their mutual interaction is a presumption for university important issues that concern improving the work of a university lecturer in agricultural education. The main interest of the authors is focused on developing a teaching style to prepare students for continuing learning in a learning society.resulting in new pedagogical work and followed by the extension of research-based knowledge to professional and vocational practice. Between the three activities of research and teaching and extension, the priorities and distribution of particular qualification. An absence of even one item changes the native of the university.

The development of their own science discipline constitutes for every university lecturer a basis for individual reputation and recognition. Moreover, it is both subjectively and objectively recognised as evidence of competency. Another situation emerges in the quality of pedagogical competencies.

The research reported here was focused on:
1. Pedagogical qualification.
2. Developing the competencies of university lecturers as a part of a complex development of a university lecturer.
3. Experience of realisation of a course for university lecturers
4. A small-scale study.

The conclusions were that, providing that by competence we understand fitness or capability to practise certain activities with professional confidence, then the course to develop pedagogical competencies had helped to develop abilities to communicate and transfer knowledge between a university lecturer and student. The aim was to develop in young lecturers a certain level of self-reflection and provoke thoughts about the ways of passing knowledge, and creating conditions for developing the skills and the study habits of the students. Together with comments their expertness they sometimes expressed partly critical remarks which mostly lacked reflection: *"and what about a student?". Is he going to understand me? What is the best way to catch his attention? If he had not understood me, wasn't it my fault? How shall I teach next time?"*. A creative and responsible lecturer should not put these and many other questions simply in order to doubt himself, but through well considered feed-back to achieve a higher level of self-confidence, and a feeling of inner satisfaction that he knows how and why he has transferred his knowledge.

The participants' responses imply that some well-balanced criticism has been generated, an appreciation developed of possible shortcomings and problems and especially the techniques of how to solve learned difficulties were discovered. A potency to be a good teacher is an inborn matter, but let there be a chance to achieve pedagogical mastery not only through experience and empiricism, but through formal pedagogical training.

References:
Blum, A. 1996. *Teaching and Learning in Agriculture*, FAO Rome.
Level, D. A., Galle, W. P., 1988. *Managerial Communications*. Homewood, Illinois 60430, USA.
Slavík, M., Miller, I., 1999. *Does a university lecturer need a pedagogical training?* In: *Proceedings published from international research seminar organised on occasion of 35th anniversary of establishing the Department of Education and Psychology* in Nitra, Slovak Agriculture University in Nitra, 114-118. (ISBN80-7137-582-9).
Slavík, M., Miškovský, Z., 1994. *On the Pedagogical skills of Young University Teachers.* Proceeding of the European Conference on Higher Agricultural Education, Wageningen Agr. University, The Netherlands, 57-61.
Wentling, T., 1993. *Planning for Effective Training*. FAO, Rome.

Agricultural Education in Transition: the Role of Educational Research

Varea, A.

Department of Clinical Veterinary Medicine, University of Cambridge, Cambridge CB3 0ES, UK

The relevance of higher education is judged by how effectively institutions adapt in order to meet the demands of stakeholders. With change being a series of more or less continuous adaptations of varying magnitude, relevance is translated into a nest of performance objectives, each contributing to, and shaped by, a culture of accountability (Gibbons, 1998). A major challenge for agricultural education is the shift in focus in the agricultural sector from production agriculture to the complexity of rural development and sustainable agriculture – defined in terms of economic, environmental and social sustainability (OECD, 2000). Much adjustment in agricultural education has been an incremental, *ad hoc* reaction to a succession of crises, and has resulted in a loss of direction and self-confidence in the sector (Warren, 1998). The present paper argues that educational research has an important role to play in the orchestration of adaptations in agricultural higher education by informing policy formation and professional practice.

Comparatively little research into agricultural higher education is published, with many papers being of a descriptive nature, focusing on the report of practices and programmes and only occasionally on their evaluation. The paper highlights the need for systematic reviews of the often-fragmented studies, and for the formulation of future research that is based on a critical examination of existing knowledge and that in turn is presented in a way that allows others to critically appraise it, thus generating a cumulative body of knowledge. Debates on the quality of educational research and its use (Davies, 1999; Kogan, 1998; Pring, 2000), and implications for research into agricultural higher education are discussed. Finally, a case is put forward for better coordination of initiatives generated by individuals, agricultural education associations, professional or profession-related bodies and informal collaboration amongst specialists across institutions.

Refererences:
Davies, P.T., 1999. What is Evidence-Based Education?. British Journal of Educational Studies, 47(2): 108-121.
Gibbons, M., 1998 Higher Education Relevance in the 21st Century. Washington, D.C.: Education, Human Development Network, World Bank.
Kogan, M.,1998. The Treatment of Research. Higher Education Quarterly, 52(1): 48-63.
　　Pring, R., 2000. Editorial: Educational Research. British Journal of Educational Studies, 48(1): 1-10.
OECD, 2000. Introduction to Issues Related to Agriculture/Environmental Interactions and Sustainable Agriculture. Proceedings of the Second Conference of Directors and Representatives of Agricultural Knowledge Systems (AKS), Agricultural Knowledge Systems Addressing Food Safety and Environmental Issues, Paris, 10-13 January 2000.
Warren, M.,1998. Practising what we preach: managing agricultural education in a changing world. Journal of Agricultural Education and Extension, 5(1):

Curriculum changes at the institutional and regional level

Contributed papers

Institutional Development Through International Programs at Moscow State Agroengineering University

Chumakov,V.[1], Bruening,T.[2]

Moscow State Agroengineering University, Russia[1]
The Pennsylvania State University, USA[2]

Since the dramatic change of the former Soviet Union, universities in Russia have been challenged to develop new curriculum and the infrastructure needed to sustain standards. In the past, universities were dependent on state resources for all operational costs. However, the economic challenges in Russia have been profound. Real per capita income in Russia is down by as much as 80%. Russia's annual government revenues are less than what the U.S. treasury collects in a single week (Zuckerman, p. 30, 1999). In agriculture, dairy herds have been reduced by over 75%. According to Borisova (1999), grain harvests are down from a high of 116.7 million tons in 1990, to 53 million tons this past year. All of these economic challenges have put greater pressure on the university to seek new ways of working and building partnerships to help redefine the university and to develop new curricular models.

In the past, state planners and developers dictated the university curriculum planning process. Now it is the role of the scientific council made up of administrators and key faculty members to set the standards for curricula development, academic rules, and general questions of academic policies at the university. Policy decisions that direct the development are discussed, debated and voted on. According to Foster (1999) there is a need for institutional change and stronger leadership for institutional reform at universities.

To foster academic and curriculum development projects, deans and administrators now have the latitude to develop innovative programs. Recently, a new proposal was developed at Moscow State Agroengineering University (MSAU) to develop a new faculty in International Agriculture. The goal of this program was multifaceted. First of all, it was theorised that this new faculty could help the university make contacts with international universities and learn different curriculum strategies, approaches, and methods. It was also believed that this new faculty could enhance the university's ability to develop collaborative projects and thus strengthen the infrastructure of the university. Finally, through increased linkages, students and professors would have greater opportunities to learn new technology, incorporate international standards, and develop a greater global perspective to operate in the international marketplace.

This paper will describe three initiatives that have been developed at Moscow State Agroengineering University as a result of developing this new international faculty:

1. Description of MSAU
2. Impetus for the Common Education Program
3. The Moscow State Agroengineering Perspective.

Methods

The methods used to create these new initiatives were participatory and developmental in nature. The two key universities, MSAU and PSU, contributed to the development of the curricula. Efforts were made by both institutions to seek solutions to problems and challenges through open dialog through email. Both universities modified courses and approaches to meet the needs of the collaborative effort. From a MSAU perspective, less technical engineering courses were offered. From a PSU standpoint, more Russian and international content examples were introduced into the curricula. The effort to create the international faculty as MSAU to accommodate the courses was challenging and successful. Additionally, at Penn State, it was a challenge to select the appropriate students to participate and recruit in the program.

Both universities sought partial support and inclusion to bring into the international program. This model begins to fit the FAO suggested approach of participatory curricular development. Students, parents, professors and industry were involved in the curricula.

Results included:-

1. Joint International Activities
2. Analysis of the International Activities
3. Implications for Educational Practice.

The conclusions were that joint international agricultural educational programs should be strengthened and enhanced across the world. There are many benefits for students, professors, and institutions through co-operative international educational endeavours as described in this paper. More European, Russian and U.S. universities should participate in future common education programs, joint curricula development activities and seminars. The programs presented here are experiments that have been successful beyond the designs imagined in the original proposals. Student and faculty exchanges are excellent ways to develop dialog and common understanding. Much was learned about the strengths and challenges of our individual university curricula through the seminars. The common education model presented here represents a new design for U.S. study abroard programs. Professors and students alike benefited in this program. If this model is developed appropriately, it could enable students to graduate on time and to make timely progress in their academic career. This model helps professors by facilitating their intensive involvement over the one-month teaching period. Students gained a great cultural understanding. New information regarding educational practice and agriculture were learned. Most U.S. professors could not expect to participate in international teaching activities for longer periods of time. This program should move forward and should be expanded to include more universities and also continue to expand the curricular development opportunities.

References

Bruening, T.H., Moran, M., and Averianova, O., (1999). *Evaluation of international study abroad program.* NAREC Publication, P. 346.

Borisova, Y. (November, 1999). *Russia's farms bring in the harvest without reaping the rewards.* St. Petersburg Times #514.

Foster, R.M. (1999). *From local to global: The challenge of change in agriculture and the food system. Leadership for Higher Education in Agriculture: Proceeding of the inaugural conference of the global consortium of higher education and research for agriculture.* Amsterdam, Netherlands.

Zuckerman, M.B. (1999). *Proud Russia on its knees: Will the West come to the rescue?* U.S. News & World Report, 2/8/99.

University integration : the most up to date challenge for agricultural higher education

Fuleky, G.

Szent Istvan University, Gödöllö, Hungary

The lecture will discuss the following questions:

1. The importance of agriculture in Hungary, before and after political change.
2. The short history of agricultural higher education in Hungary.
3. Changes in the curriculum of agricultural higher education in recent decades.
4. New challenges:
 - Privatisation
 - transition of co-operative farms and state farms
 - EU integration possibilities,
 - university integration,
 - parallel development of agriculture and rural areas.

The present situation with be discussed using the example of the new Szent Istvan University Godollo.

Evaluating the Guelph Electronic MBA in Agriculture: Next Steps Globalisation

Parton, K.A., Funk, T.F.

Department of Agricultural Economics and Business, University of Guelph, Guelph ON N1G 2W1, Canada

In January 1997, the MBA in Agriculture at the University of Guelph admitted its first students to class. This was pioneering stuff, with the main mode of delivery being computer-based electronic communication to and among students spread across Canada and the U.S. Nowhere else in North America had University administrators been bold enough to back this emerging technology in a program specialised to agriculture. Now with more than three years of experience in running the program, it is time to present an assessment of the approach. To say that the outcome has been most favourable would not be an exaggeration, but nevertheless many unexpected events happened along the way. The paper presents a systematic analysis of the program that was developed emphasising the advantages and disadvantages of the delivery mode. The delivery method is first described, because how things come together is usually only vaguely understood by those outside such a program. In this context, the asynchronous communication method is a central issue. The remainder of the paper is concerned with evaluation of the program, based on in-depth interviews with both students and staff. The final section suggests appropriate changes for the future, given the way in which both the technology and the client base in agribusiness is evolving.

Economics of Mountain Systems - a network of European universities training regional specialists for the European mountain areas

Meixner, O.

Institute of Agricultural Economics, University of Agricultural Sciences, Peter Jordan Str. 82, A-1190 Vienna, Austria.

The following report gives a general picture of the CDA research project "The Economics of Mountain Systems" which is being supported by the European Union as part of the SOCRATES / ERASMUS programme (CDA = curriculum development – advanced level).

The main aim of the project "Economics of Mountain Systems" is to create a course which enables the education of regional experts and give the students the skills and knowledge they need if they are to contribute to the regional development of Europe's mountain areas. These experts should be prepared to help the native population with specific economic, ecological and social problems arising from the regions' geographic location. Alongside the development of the curriculum there are further, lateral aims: the establishment of a functional network, made possible above all by ITC media (information and telecommunication media: email, internet, video conferencing etc.); the preparation of teaching materials for open distance learning (ODL); preparation for legal implementation of the course in each partner state; preparation of the necessary examination system for the mutual recognition of academic achievements as part of the European Credit Transfer System (ECTS). Organisational issues are also being resolved so that the actual course can be implemented as smoothly as possible. Right from the start the aim has been to create a teaching programme based on ODL, so the course can be offered jointly at all partner universities. Such an approach is only possible if the course takes a modular format. Five such modules are planned (see diagram 1).

Module		responsible Partner
M1	Human Resource Management	Austria University of Agricultural Sciences Vienna
	Thesis [1]	
M2	Agriculture & Forestry	Italy Università Degli Studi di Milano
	Thesis	
M3	Regional Marketing	Germany Technical University of Munich
	Thesis	
M4	Diversivication of Mountain Economic Systems	Portugal Universidade de Trás-os-Montes e Alto Douro
	Thesis	
M5	Regional Development	France Pr Economie et Politique Agricoles, ISARA
	Thesis	

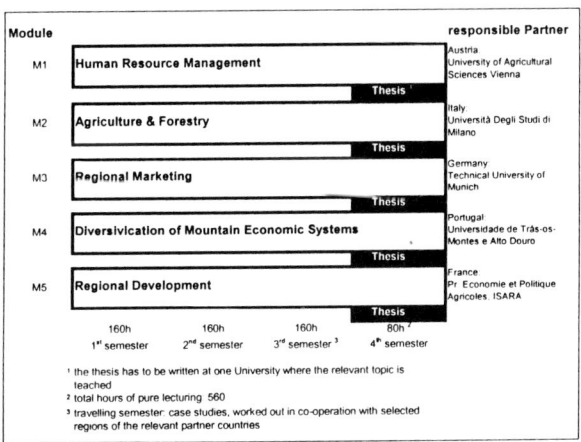

160h 1st semester   160h 2nd semester   160h 3rd semester [3]   80h [2] 4th semester

[1] the thesis has to be written at one University where the relevant topic is teached
[2] total hours of pure lecturing: 560
[3] travelling semester: case studies, worked out in co-operation with selected regions of the relevant partner countries

Diagram 1: preliminary schedule of the EMS course

Each project partner is responsible for developing one of the modules (see diagram 2 for module 1) and each module will only be incorporated into the in situ teaching programme of the University responsible for its development (and offered to the other four universities via ODL). This course structure means the course will be both cost effective and truly international in perspective. The network of partner universities is made up of representatives from Austria, Germany, Italy, France and Portugal.

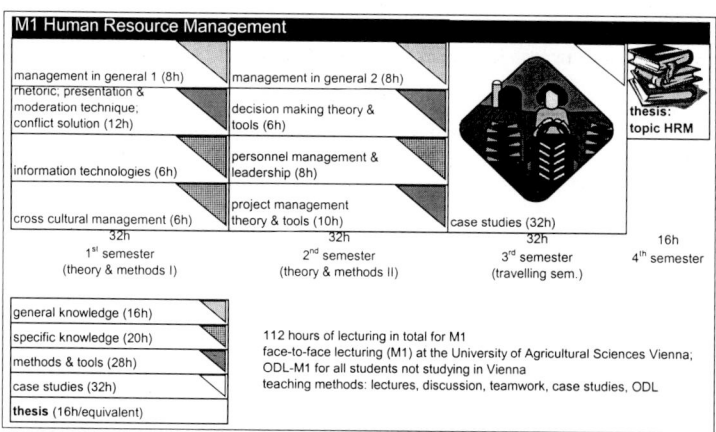

```
┌─────────────────────────────────────────────────────────────────────────────┐
│ M1 Human Resource Management                                                │
│                                                                             │
│ management in general 1 (8h)    management in general 2 (8h)                │
│ rhetoric; presentation &                                                    │
│ moderation technique;           decision making theory &        thesis:     │
│ conflict solution (12h)         tools (6h)                      topic HRM   │
│                                                                             │
│                                 personnel management &                      │
│ information technologies (6h)   leadership (8h)                             │
│                                                                             │
│                                 project management                          │
│ cross cultural management (6h)  theory & tools (10h)   case studies (32h)   │
│         32h                            32h                  32h        16h  │
│     1st semester                   2nd semester        3rd semester  4th semester │
│   (theory & methods I)           (theory & methods II)  (travelling sem.)   │
│                                                                             │
│ general knowledge (16h)                                                     │
│ specific knowledge (20h)         112 hours of lecturing in total for M1     │
│ methods & tools (28h)            face-to-face lecturing (M1) at the University of Agricultural Sciences Vienna; │
│                                  ODL-M1 for all students not studying in Vienna │
│ case studies (32h)               teaching methods: lectures, discussion, teamwork, case studies, ODL │
│ thesis (16h/equivalent)                                                     │
└─────────────────────────────────────────────────────────────────────────────┘
```

Diagram 2: course schedule of M1

The development of this *interdisciplinary* teaching programme will be finished late 2000. Afterwards, the course should be introduced on a post-graduate level (Master) at the universities of the project partners. Five European universities have been co-operating on this project since 1997/1998, the aim being to develop an appropriate curriculum and the associated teaching contents and teaching materials.

The EMS network:
- Austria: University of Agricultural Sciences Vienna (co-ordinator)
- Italy: Università Degli Studi di Milano
- Germany: Technical University of Munich
- Portugal: Universidade de Trás-os-Montes e Alto Douro
- France: Pr. Economie et Politique Agricoles, ISARA

Spain, Switzerland, the Balkans, Greece, Eastern Europe and all of Scandinavia represent obvious "gaps" in view of European mountain regions. However, the five partner Universities that were able to take part in the EMS project have enabled an effective structure which is worthy of the description "European" and which is worthy of the description "modern higher agricultural education" as well.

The experience of the Buryat State Agricultural Academy in improving ecological education in training students of agriculture

Popov,A.P., Korsunova,T.M., Radnotarov.V.D.

Buryat State Agriculutral Academy named afetr Philippov,V.R., PPushkina 8, Ulan-Ude, Russia 670024.

It is possible to realize the strategy of ecologically sustainable agriculture in the Baikal region , to make agricultural production ecologically friendlier by paying more attention to ecological education in training specialists at higher educational establishments, colleges, by introducing ecological education into the whole process of training. Today, an agricultural specialists must be provided with systematized professional knowledge of ecologically friendly agricultural production process in accordance with ecological laws of farming, methods of neutralization and elimination dangerous wastes, ecological optimization of agrolandscapes. Ecological aspects, priorities of ecologically friendly agricultural production with regard to peculiarities of the region agroecosystems and necessities to increase agrosystem sustainability and stability should be included into all courses. Great attention should be paid to the consideration of economic mechanisms of nature utilization rationalization under market economy, to studying the bases of ecological law.

The Buryat State Agricultural Academy realizes it thorough ecological training of specialists, by continuous ecological education through complex of general, special subjects and extracurricular courses, through the whole educational process, by carrying out ecological scientific researches.

Thus at all faculties both full-time and extra- mural students of all specialities are taught special courses such as: "The bases of ecology and environmental protection", where questions of biosphere ecology, the peculiarities of ecological laws realization in reference to agrarian sphere, conservation of natural recourses are considered. Besides, there is a special courses for students of economics faculty: "The economy of nature utilization", where economic methods of rationalization of nature utilization by agriculture, estimation of ecological and economic damage as a result of environmental pollution and discharges from certain industrial enterprises are considered.

In order to improve ecological education at the agronomy faculty there have been introduced a new speciality "Agroecology" to consider thoroughly problems of ecology and agronomy, ecological law of farming and to train specialists majoring in ecological optimization of agrolandscape and producing ecologically pure farming production, organizing agroecological monitoring.

A new paradigm of ecological education, being realized in the Baikal region, is based on intensification national region component, realization of the principle: "to think globally, to act locally". There have been realized region approaches based on considering the peculiarities of the region ecosystems, social economic causes of ecological tension and new approach to solve local ecological economic tasks, united into the strategy of sustainable development.

To make ecological education successful it is important to take into consideration social- ethnic and national peculiarities of the region inhabitants, to study examples and experience of interaction between nature and people, living in the Baikal region, particularly the Buryat people experience in environmental protection, the Buddhist main principles as regard to the necessity of individual spiritual improvement and needs limitation. To use these examples in teaching students makes it possible to bring up ecologically mineded people and realize the strategy of sustainable development.

The approaches have been realized by elaborating and introducing new courses, adapted to the peculiarities of the region: "Ecological economy in the Baikal region", "Ecological problems in the Baikal region", "Problems and perspectives of sustainable development of the Baikal region", "The present environmental situation in rural locality" and number of other optional courses, stating ecological problems and suggesting different approached to solve them.

In a whole, the global ecological problems discussed within courses are given as regional problems, typical for a certain agricultural enterprise and locality.

Practical training on ecology, provided by the curricula, makes possible to study the peculiarities of the regional nature and agrosystems, estimate forms and consequences of antropogenic effect. There have been worked out ecological tasks for final research practical work of graduates, which aims at data collecting on ecological situation of natural resources and objects, registering factors of nature protection violation in the process of agricultural production and estimating ecological situation.

It is obligatory for all graduates to include the part devoted to environmental protection into the final research work with ecological analyses of a certain farm, scheme of present situation of natural recourses and recommendations on improving environmental protection on the farm, enterprise.

Students ecological contests and scientific conferences on ecology are held every year.

The improving of ecological education in training specialists for agriculture, introducing ecological education into the whole process of education contribute to the effective cropgrowing, which is able to provide population with ecologically sound agricultural products and sustainable development of the region.

Agricultural University of Poznań Poland - Curriculum changes in last decade and new ideas for higher education in future

Skrzypczak, G., Drzymała, S.

Agriculture University of Poznań, Faculty of Agronomy, 60-637 Poznań, Poland

The last decades in Poland are characterised by significant – sometimes "revolutionary" – changes in many branches, also in higher education. In the academic year of 1998/99 the number of universities and colleges in Poland was about 142 more than in 1991/92. This very significant increasing of number of universities and colleges is caused by new Polish law, which (since 1991), allow to established private schools and colleges. Almost all (except 2) new colleges are private and most of them are on "economic field", but only one private College of Agriculture was created during last decade. At the same time, number of students also significantly increased (about 160%) especially "post-diploma" and PhD students. At agriculture universities at the beginning of 1990's were learned about 41 000 students – that was 7.6% of all students in Poland, but in 1999 amount of students increased more than two times (85 000) but it was only 6.7% of all students.

The ratio of students of typical agriculture faculties (agronomy, animal husbandry, horticulture) is much lower and amounts to 3.1%. These "kind" of students among all students on agricultural universities on 1990's was 70% but on 1999 – only 45%. It means that at present more than 50% of all students on agriculture universities are studying other specialisations than strictly connected with agriculture production. In the table 1 are shown most important changes in higher agriculture education in respect of general changes. These general trends are also observed in the August Cieszkowski Agriculture University of Poznań – one of the biggest Agriculture Universities in Poland.

Table 1. Most important changes in higher education in Poland in the last decade.

Number of	1992			1999		
	Total	Public	Private	Total	Public	Private
All Universities and Colleges	124	106	18	266	108	158
Agriculture Universities and Colleges	9	9	0	10	9	1
All Universities and College Students*	495.7	479.5	16.2	1274.0	942.5	331.5
Students of Agricultural Universities*	40.9	40.9	0	85.5	84.6	0.9
All "post-diploma" students*	10.9	10.9	0	61.1	61.1	0
Agriculture "post-diploma" students*	0.84	0.84	0	6.5	6.5	0
All PhD students*	2.08	2.08	0	18.9	18.9	0
Agriculture PhD Students*	0.11	0.11	0	2.0	2.0	0

* - in thousands

In the last decade Polish agriculture has been drastically changed. The most of the economically non-effective state and co-operative farms – for which our higher education was mainly adopted and where our students had practice training – had crashed. This fact was one of the main reasons of significant changes in curricula of most faculties of our University especially on biggest and oldest (created in 1919) Faculty of Agronomy.

The new curriculum is oriented not only for production but includes subjects for rural development, social activity and life in rural areas. The new curriculum give also much more opportunity for study of economic and marketing aspects in agriculture and food technology. The obligatory training – on best Polish private and new "state" – farms and on different farms in many western countries – play very important role in our education system.

The most important changes in last decade in our University education system are establishing several new courses. On beginning of 1990's our University comprise on 7 faculties, which offered only 10 different courses (specialisations). At present the number of faculties is the same (7) but we offer 21 different specialisations. For example the Faculty of Agronomy on 1990's offered only 2 specialisations (agriculture and agriculture-engineering), but at present we offer 9 specialisations: agronomy (with 7 sub-specialisations), plant protection, genetic and seeds production, agriculture engineering, economics of agriculture engineering, informatics and computer sciences in agriculture engineering, biotechnology, economics of food production and environmental protection. The last three courses are most preferable by students (4-6 candidates for 1 place). The agronomy and engineering in agriculture besides very big changes of their curricula are still much less interested. The number of students on these 2 specialisations is only about ¼ of all students of Faculty of Agronomy and show rather decreasing trend.

The most important changes in curricula are as follow:
1. introducing "facultative subjects" – which are chosen by students from very broad list of subjects. These subjects comprise about 25% of all;
2. obligatory subjects (75%) are classified into 3 groups:
- "general subjects" (significant changes): introducing to every courses, mathematical statistics, informatics, computers and at least one foreign western language,
- "professional subject" (relatively small changes): almost all subjects are based on lectures and laboratory exercise,
- "economical and law subjects" (the biggest changes): introducing such subjects as marketing financing, rural development, rural sociology, agriculture-policy, EC-law and economy etc.

The main goals of our future education will focus on adaptation of curricula for further transition in agriculture in our country and "smooth" incorporation of Polish agriculture to European Community in near future.

References:
Drzymała, S., 1999. The role of training in improving agricultural education (in Polish). Problemy dydaktyki i wychowania w Akademii Rolniczej w Poznaniu. Wydawnictwo AR Poznań Nr 15: 39-43.
Skrzypczak, G., Wiatrak, A. W., 1997. Agriculture extension during transition to a market economy. Yale-Tübingen Summer Seminar – Occasional Paper No. 4. Economic Growth Center – Yale University, New Haven, USA: II 4-7.
Skrzypczak, G., 1997. The situation and outlook for manufactured agricultural inputs in Poland – outlook report as the extension information. Yale-Tübingen Summer Seminar– Occasional Paper No. 4. Economic Growth Center – Yale University, New Haven, USA: II 29-31.
Yearly statistic books of Poland. GUS Warszawa 1991 and 1999.

Agricultural to rural development education in practice

Contributed papers

From Reflection to Action: Higher Education in Rural Development at the University of Cordoba

Ramos, E., del Mar Delgado, M

Rural Development Team, Universidad de Córdoba, 14080 Córdoba, Spain.

This paper is structured in 2 sections: the first one details the internal and external criteria considered to design and commission a program of high level courses in Rural Development in a Spanish University long specialised in higher agriculture education. The second section describes the academic curriculum and the methodology used.

The handicaps associated with the change from a traditionally considered production agriculture to an emerging approach in Rural Development courses have been evaluated. Any negative factor depends equally on both the offer and the demand for new courses in Rural Development. The main cause of the lack of offer and demand in Rural Development is the strong inertia of the past, affecting both, the economical and social agents of the sector and the academic staff.

The needs of highly qualified professionals were the answer to overcome to double limitations above. The profile of the new professionals was defined. From these profiles, the need of higher education in Rural Development emerged. Innovation and flexibility within the nature of content of the program were essential factors, mainly in the definition of the theoretical paradigm, subject contents, tools and methodologies.

The first handicap to overcome was the lack of an implemented paradigm on the subject definition. The concept of Rural Development itself and the debate on how to accurately define it, took a significant role in the initial reflection. The experience gained by the leading team in a regional debate on Rural Development and a public survey carried through different public and private agents proved to be paramount. Besides, it gave credibility and social support to the process.

The diversity of subjects that fulfil the new professional needs can originate clashes with certain traditional courses. This factor proves that there are important holes and lack of integrative visions with a regional focus. Flexibility and diversity were the key words to define subjects combining theoretical contents and practical tools. The concepts of 'bottom up' and 'learning by doing' are essential to the program.

The second section describes the commissioning process of the first program of higher education in Rural Development in Spain, initiated in 1995. The offer of the University of Cordoba follows on the conclusions of the reflection detailed above and it continuously adjusts itself to satisfy current demands. The *Titulación Superior en Desarrollo Rural-TSDR* (High Degree in Rural Development) started in 1995. Additional products as the *Master Internacional en Desarrollo Rural- MIDR* (MSc International Rural Development) or the *Master en Gestión del Desarrollo Rural-MGDR* (MSc Rural Development Management) complete the offer.

A distance learning program is being developed to satisfy the needs of those professionals who cannot attend a full time course in Southern of Spain.

The connection between these academic offers and public and private agents, related with the issue, has always been a key concern in the definition and development of this education system.

Interdisciplinarity, internationalisation and learning systems for agricultural education in transition.

Gibbon, D., Jiggins, J.

Department of Rural Development Studies, Swedish University of Agricultural Sciences, Uppsala, Sweden.

Many agricultural higher education institutions in Europe, which have concentrated for the past 50 years on training agriculturalists and agricultural scientists to work in support of a modernising, capital intensive agriculture, are now engaged in fundamental debates about future directions, strategies, curricula and modes of learning.

The training that has been developed over this period has been, by and large, dominated by positivist/reductionist production oriented science and institutional structures which have been increasingly fragmented through the formation of commodity and discipline-based departments. Admittedly the disciplines bring depth, but they also confer "territories" and the defence of boundaries. The orthodoxy stifles innovation in learning styles, particularly at undergraduate level.

The output from this kind of agricultural training has failed to address the complexities of adaptive human/natural resource management and rural/urban livelihood systems that are now common in many rural and rural /urban systems throughout the world. The reasons for this state of affairs are complex and strongly connected with the role of science-based education and the historical role of agriculture as the primary producer of food for many nation states. Also, policies have been introduced which have supported the idea that agricultural research and extension should focus on yield maximisation, market driven efficiency and subsidy and protection, in a variety of ways, by the state.

Changing roles and involvement of the state, health and environmental concerns, both in the North and South, a widening interest in more sustainable natural resource management and rural livelihood systems and declining student numbers applying for places in conventional agriculture courses (at Reading, the Swedish Agricultural University and Wageningen for example) , prompt a need to rethink the mode and form of training that students receive at agricultural higher education institutions. This paper discusses some alternative approaches, including stronger international linkages and learning, the structure and organisation of agricultural universities and the style and content of training and learning programmes. The key areas of discussion revolve around; interdisciplinarity, systems thinking, participatory action and shared experiential learning. The paper acknowledges that experiments in reform of this kind have been initiated before and have not always been sustainable. Some possible explanations for this are indicated and alternative strategies that might prove to be more sustainable, are discussed. The challenge, if agricultural universities and faculties are to have a future, is to generate self organising and managing, reflective, educational systems which respond to society's present and future needs with respect to the management of natural resources and land-based production systems within rural and rural/urban livelihood systems.

Sustainable Rural Development and Curricula in Higher Educational Institutes: Issues and Challenges

Koutsouris, A.

*Department of Rural Economics & Development, Agricultural University of Athens
118-55 Athens, Greece.*

Education is widely accepted as being a key policy instrument for bringing about a transition towards a sustainable future. In parallel, education has itself to be transformed. Therefore, education will simultaneously be a change agent and a subject of change; currently, education is part of both the problem and the solution.

Sustainability as well as (sustainable/ integrated) rural development call for the close examination of the dynamic balance among many factors such as political, technological, economic, socio-cultural and environmental. A resolution thus to the current strategic crisis is not primarily a technical problem; it requires a fundamental change of existing social structures.

Curricular studies, in their broadest sense, contain quite powerful debates on the nature of objectives, knowledge, methodologies and evaluation. Such debates seem to have not seriously affected (mainstream) education till recently. The productivist approaches of the present curricula have not allowed alternative pedagogical arguments to penetrate the discourse of educational institutions. The issue of sustainability and more recently of rural development contribute to the animation of a lively dialogue on approaches to education, at least during the last couple of decades. Nevertheless, most of the institutions are still not affected by such concerns.

Like sustainability *per se*, education for sustainability (EFS) has a range of meanings in theory and practice, thus not escaping from philosophical and political debates. The present paper intends to discuss issues related to curricular studies *vis a vis* sustainable rural development. Through the use of concepts such as farming and soft systems a critical understanding of curricula in higher agricultural education will be undertaken.

Thus, changes towards alternative curricula will be discussed. Such changes include the abandonment of narrow, disciplinary knowledge, one-way/ hierarchical, stimulus-response teaching and students' examination as evaluation. Instead, it is argued, systemic/ integrated curricula should develop; interactive strategies and theme-based experiential methods i.e. the dynamic interaction of action and reflection within real learning situations should be promoted; then, learning should be understood as a social act and judgements about the quality and meaningfulness of the work should be controlled by the participants in the learning situation. Action-research, critical pedagogy and praxis are the main concepts/ practices within a new educational paradigm aiming at sustainability.

References:
Brown, V., Smith, D.J., Wiseman, R. &. Handmer,J. 1995. Risks and opportunities: managing environmental conflict and change, Earthscan Publ. Ltd., London, UK.
Checkland, P. and J. Scholes, 1999. Soft Systems Methodology in Action, John Wiley and Sons, Ltd., Chichester, UK.
Huckle, J. and S. Sterling (eds.), 1996. Education for Sustainability, Earthscan Publ. Ltd., London, UK.

Ison, R. and D. Russel (eds.), 1999 Agricultural Extension and Rural Development: Breaking out of traditions, University Press, Cambridge, UK.

McKernan, J., 1991. Curriculum Action Research, Kogan Page Ltd, London, UK.

Norgaard, R., 1994. Development Betrayed: the end of progress and a coevolutionary revisioning of the future, Routledge, London, UK.

Roling, N and M. Wagemakers (eds.), 1998. Facilitating Sustainable Agriculture, Cambridge University Press, Cambridge, UK.

Shepherd, A., 1998. Sustainable Rural Development, MacMillan Press Ltd., London, UK.

UNESCO, 1997. Educating for a Sustainable Future: A Transdisciplinary Vision for Concerted Action, Paris, France.

Young, M. F. D. (ed.), 1971. Knowledge and Control, Collier-McMillan Publ., London, UK.

Graduate Education via the Internet: A World Campus Graduate Certificate in Community and Economic Development (CEDEV. Certificate)

Hyman, D., Holsing, C.

The Pennsylvania State University Pennsylvania, USA

The Department of Agricultural Economics and Rural Sociology at The Pennsylvania State University (Penn State), USA, is partnering with the Penn State World Campus to create a graduate certificate program in Community and Economic Development (CEDev). The program was formally approved in May 2000 and will be offered in conjunction with an Master of Science (MS) degree, also titled Community and Economic Development. The electronic versions of five core courses required for the MS will be offered via the Internet in phased order beginning in January 2001.

Internationally, agricultural higher education is presented with unprecedented opportunities to reach out to communities, organisations, governments, and individuals to educate and inform an ever larger number of citizens to address the issues of development and underdevelopment, industrialisation and growth, suburbanisation and sprawl, equality and democracy, and creation of sustainable communities. It seems appropriate at this nascent stage of electronic education to enter a conversation about problems and prospects of graduate education in community and economic development over the Internet. This paper presents an overview of the CEDev Graduate Certificate as a basis for engaging colleagues in this conversation.

Sustainability of Rural Areas in Transition and Higher Agricultural Education

Rivža, B., Krūzmētra, M.

Latvia University of Agriculture, Jelgava, Latvia

The 90ties is time in Latvia, when social economical system is changing in all areas of society and all territorial units. They are concerning cities and countryside, production and consumption, institute of state administration and inhabitants etc. These changes concern labour market and education, what are linked apparitions and in this case the object of analysis. On the one hand, labour market requires to higher educational institutions to prepare specialists, adequate to market needs, on the other hand, higher educational institutions influence labour market by preparing and turning out in labour market definite number of definite professions people. For successful development of society the balance of these two society institutions and coherence of action is necessary.

It is possible to divide in the view of interconnection of these two systems significant changes in countryside and higher agricultural education:

Changes in Latvia's countryside:
- Re-established private property to land, wherewith small manufacturing is form up in remarkable measure;
- The loss of the Russian market is wherewith minimised demands for agricultural production;
- The wind-down of agricultural production on the one hand calls in a part of housekeeping the transition to natural economy, and on the other hand searching other occupation versions;
- Territorial structures became determinant in the place of manufacturing structures.

Changes in the higher agricultural education:
- The action's wind-down of few faculties there of demands of labour market wind-down in specialities like hydromelioration, agricultural mechanisation and zootechnics;
- The action's enlargement of few faculties thereof demands of labour market increase in specialities like economics, land use planning;
- The opening of new programmes in specialities like entrepreneurship, country development, country tourism, the sociology of organisations and public authority;
- Work rearrangement in all specialities: the trimming of study content to existing needs of life and European standards.

The introduction of new study programmes is easier than modification of existing study programmes. The last asks for break the tradition, what in many cases makes not only organisational but also psychological difficulties.

The new development paradigm (J. E. Stiglitz, 1998) understands development as constant transformation of society. Development represents a transformation of society, a movement from traditional relations, traditional ways of thinking, traditional ways of dealing with health and education, traditional methods of production, to more "modern" ways. This access lays to higher educational institutions be more flexible, in time perceive passing changes in countryside society, to make modifications in the study programmes, but better - analysing processes of country development forecast development perspectives, through specialists prepared by new study programmes involve development of desirable way in country development thus increasing the influence of education to process of development of society.

References.
Rural Development. Within the Process of Integration into the European Union. International scientific conference reports. - Jelgava, 2000.
Co-operation and country problems. The example of Latvia Rietumvidzeme region. Latvian State Institute of Agrarian Economics. Working Papers. 5/1999.-Riga, 1999.
Latvia University of Agriculture. Data of study research. 1998./99. Academical year.- Jelgava, 1999.
The National Concept of the Development of Higher Education and institutions of Higher Education of the Republic of Latvia.- Riga, 1998.
Stiglitz. J.E. Towards a New Paradigm for Development: Strategies, Policies and Processes. (1998) Publish Lecture at UNCTAD, Geneva, October 19, 1998.
Teaching and learning in Agriculture. A guide for agricultural educations- Rome, FAO, 1996.
State Higher Education institutions in Latvia.- Riga, 1996.
Latvia University of Agriculture. Faculty of Economics. Documents for the accreditation to the study program in Economics.- Jelgava, 1996.
Ministry of Economy. Republic of Latvia. Economic Development of Latvia. Report. Riga, 1993.
A European agenda for change for Higher education in the XXI century. UNESCO, 1997.
Issues and opportunities for Agricultural Education and Training in the 1990s and beyond.- Rome, FAO, 1997.

Educational Innovations Workshop

Contributed papers

Producing Quality Video for Use as a Teaching Aide

Polson, J.

The Ohio State University, Wooster, Ohio 4469, U.S.A.

This paper offers guidelines based on personal experience, the literature, and suggestions from professional videographers to help the agricultural educator decide whether or not to attempt personal video production for educational purposes.

Video is a powerful teaching tool that combines motion, sound, context and emotion in ways not possible with more traditional teaching tools. Agricultural educators frequently visit farms, conduct personal interviews, witness exceptional interpersonal interchange and participate in other activities that would greatly benefit their students. Video is arguably the "next best thing" to being there when time, distance, and the large number of students make it impossible for students to participate. A well-produced video can bring large numbers of students in contact with innovative leaders and masters in various fields.

I have on occasion become a cameraman, editor, and video producer for educational purposes. I select very few of the subjects I tape for editing and use with students. I use the videos I produce in several ways to help teach adult farmers.

Recent advances in computer and video technology, and sharply dropping prices for both, have made it much easier and more affordable for educators to capture and produce high quality educational video. In 1999, Apple, Compaq and Sony each introduced new computers priced less than £1,000 that are designed to edit video. These low-priced editing systems can capture, edit and output high-quality digital video suitable for use in agricultural education.

These recent computer introductions are potentially important to agricultural educators in at least four ways. One, £1,000 is half or less the cost of most comparable systems a year or so ago. More educators can now afford a video editing system. Two, the video editing systems are ready to use when they arrive. Before these introductions, persons wanting to edit video generally found it much less expensive to purchase a video capture card and editing software from one company, a large ultra-wide high-speed disk drive from another, and perhaps some additional computer ram for their existing computer. It was easy to spend £1,000 just for add-ons. Three, the new low-priced, ready-to-edit systems greatly reduce problems with installation and compatibility. It frequently is not easy or quick to install a capture card and high-speed disk drive into a computer and get them and the video editing software to work together properly. Some have learned the hard way that it is not possible to retrofit some existing computers for video editing. Four, some new video editing software included with these computers is much simpler, easier to use and faster than the software available a few years ago. However, it still is not easy to learn some of the more powerful video editing software. Educators frequently will benefit from software training beyond the training manual.

Other notable recent technological advances include "affordable" large high-speed disk drives. Video capture cards and editing software are becoming more reliable, easier to use, and more compatible with other components in the system. Computers with more powerful and faster CPU's and more RAM make the systems more reliable and faster. Digital camcorders with a FireWire port can directly feed a digital signal into a computer with little loss of quality, if the editing system has a compatible FireWire port.

Arguably more important than hardware and software are the video producer's organisational ability, patience, skill, experience, creativity, and time. Video production requires careful planning and a substantial commitment, especially when the educator is working alone.

Time is the largest cost of most video production. Most agricultural educators will not want to or be able to commit the time it takes to choose a video production system, learn to use it, tape the footage and produce the final video. Video production is generally only justified when the resulting video is likely to have a significant educational impact and the number of potential viewers is large or influential.

Once a system is up and running and the user has learned to use the editing software, he/she should expect to spend an hour or more editing for every minute in the final video. If it takes less time the video is probably made up of a series of "talking heads," which is boring to watch. Producing an engaging educational video lasting more than a few minutes usually requires considerably more than simply splicing together different shots.

In my experience, the minimum equipment requirements for a successful video include a video camera that will accept an external microphone, earphones for monitoring audio, some means of editing, a video deck to record the final product, a subject or subjects with something worthwhile to share and extra batteries. All are indispensable!

Other potentially useful items include one or more assistants, a tripod, lights and reflectors, one or two additional kinds of external microphones, an audio mixer, a video camera that is digital, Hi8 or S-VHS, and non-linear editing equipment with one or more special large high-speed computer disks. There are many additional items others would add to the list.

Quality video production is a very complex task involving many variables, but one can learn the basic skills. Interested educators should read, study, and perhaps take a class to learn from other people's experiences. However, there is no substitute for shooting lots of video, watching it carefully, editing it, having others critique it and learning from your successes and mistakes. If possible, ask someone trained and experienced in video production to watch and critique your video. Critical review of television and movie productions is useful. Television commercials and music videos are very carefully produced. Watch them carefully, observing the timing and how the pieces fit together. Listen to how they use music to enhance the video.

In a perfect world, every student would serve an apprenticeship with innovative leaders and masters in their field learning from them while they work. In agriculture education, this innovative leader may be a farmer, agri-business person, outstanding educator or other leader. In the real world, the large number of students, the limited number and availability of innovative leaders, limitations on time, distance and finances make it impossible for every student to spend time as an apprentice with leaders in their fields. A well-produced video can help more students learn from these leaders and masters in various fields.

References:

McCleskey, Joe, Collins, Don (February 2000). An iMac That Edits DV. Computer VideoMaker, February 2000, 39-40, http://www.videomaker.com

Decker, Daniel, J. Merrill, William G. (Spring 1990). Influencing Practices through Videotape.A Systematic Evaluation of Communications Technology. Journal of Extension. v28 P21-24. Spr 1990.

Gunawardena, C. (1998). Using communications technologies for distance education: Report on current practices in the United States. (ERIC Document Reproduction Service No. ED327147)

Howes, L. & Pettengill, C. (1993, June). Designing instructional multimedia presentations: A seven-step process. T.H.E. Journal, 58-61

Ludlow, B. and Duff, M., (1997) Creating and using video segments for rural teacher education.(ERIC Document Reproduction Service No. ED 406106)

Miller, G & Honeyman, M (1993). Videotapeutilization and effective videotape instructional practices in an off-campus agriculture degree program. Journal of Agricultural Education, 35(1), 43-48

Owen, M., & Hotchkis, R. (1991). Who benefits from distance education? A study ofAthabasca University graduates, 1985-1990. (ERIC Document Reproduction Service No. ED341301)

Polson, Jim G. (1999) Using Video of a Master Farmer to Teach Others, Journal of Extension 37 (2) 33-39

Richards, T., Chignell, M.H., & Lacy, R.M. (1990). Integrating hypermedia: Bridging the missing links. Academic Computing, 4(4), 24-26, 39-44

Sauer, J. (1996). Harnessing the power of digital video. Presentations, 0(10), 57-66

Model based Online Agricultural Education for Farm Managers in a Transition Surrounding

Mothes, V.

Institute of Agricultural Development in Central and Eastern Europe (IAMO), D-06120 Hall, Germany

Introduction

In Central and Eastern Europe agriculture is on the way from political cuddling to market orientation. Farm managers have to do both to learn how market economy influences the economic success of agricultural production and to manage their small individual farm or large scale agricultural enterprise. So, they have not enough time for extensive studies. Under these conditions, the Internet enables a very convenient way to support these farm managers although they have very specific questions resulting on their daily business. Online information providers have to reflect this need.

Methodology

The paper announced includes strategies, objectives, requirements, examples and last but not least possible advantages of online agricultural education for active farmer and farm managers especially in CEECs.

Model based online education has to describe the relation between special activities planned by the farm manager and their economic effects. This relation is influenced by the factor availability in farms and specific aims (e. g. farm income or profit in agricultural enterprises). Furthermore, it is closely related to the adaptation to changes in the farm environment. Mathematical models have to reflect three main groups of indicators:

1. the processes of agricultural production, the farm environment and the rational behaviour of profit-maximising economic agents
2. experiences and customs, historical knowledge, political, juridicial and philosophical perceptions of farm manager
3. in CEECs: typical effects of transition: a high variability of the farm structure, rapid changes on the terms of trade, uncertainty about the production functions, backlog demand for investments in machinery for crop production and facilities for animal production, deficiencies in co-operation between farms, a lack of farmers' unions, a farm advisory system in construction or not existing.

Farm managers need information related to factor allocation in agricultural enterprises, investment calculation or factor application by changing terms of trade. Internet providers which dispose of suitable mathematical models to analyse these problems offer they via the Internet, e. g. with a detailed model description on their homepage.

To start the model application, the farm manager has to send primary farm data to the provider, such as :
- description of the problem,
- data determining the actual situation on the farm,
- specific goals, e. g. pure profit maximisation, job creation, farm-household consolidation,
- possible strategies to realise goals (behavioural patterns), e. g. willingness to enlarge the farm, specialisation or distribution of production.

After having received all necessary data on the provider's side starts the model-based calculation. The results are put into tables and figures to enlarge their didactic value. In this form the provider sends them back to the farm manager. This process is repeatable in different variants.

Results

The Linear Programming (LP) farm model TRANS-FARM has been especially developed for Central and Eastern European Countries. Model based information are created to evaluate cost saving cropping schedules in farms, like selection of investment strategies to mechanise crop production, prediction of effects of horizontal co-operation, assessment of the process of growing of agricultural enterprises, e. g. the extension of arable areas.

Figure 1 shows the cost-saving effect obtained by an increased use of a self propelled forage harvester: Price 100.000 GBP; Amortisation (AS) 300 ha/a; useful in time (ZG): 400 ha /a; fuel, oil, repairs (calculated): 23 GBP/ha(h). The price for hiring a server for chaffing corn with this engine: 73 GBP/ha.

In an agricultural enterprise the cost for chaffing grass depends on the decision whether this task should be done self-mechanised or hired by a server. Figure 1 shows that self-mechanisation is profitable only in a range between 270 ha and 400 ha. For an area up to 270 ha/a, a server should be hired for chaffing grass. For more than 400 ha/a should be used an own harvester and an additional hired one.

Another frequently ask question in agricultural enterprises is focused on the profitability of using own machines and own drivers in other enterprises. This question can be answered by an LP-Model providing the shadow prices of an activity. The profitability of chaffing corn with the above mentioned harvester in other enterprises depends on the market price (73 GBP) and the shadow price of this task in the own enterprise. In Figure 1, the shadow prices changed erratic from 23 GBP to 57 GBP, the profit from 50 GBP to 16 GBP, determining the kind of amortisation for taxes: in time up to 300 ha/a; in use after 300 ha/a.

References

Mothes, V. et al. (1999): An Online Information System – Core of a Research Network for Farm Development Specifically for Central and Eastern Europe (TRANS-FARM), in: Role and Potential of IT, Intranet and Internet for Advisory Services, Bonn, S. 155-164.

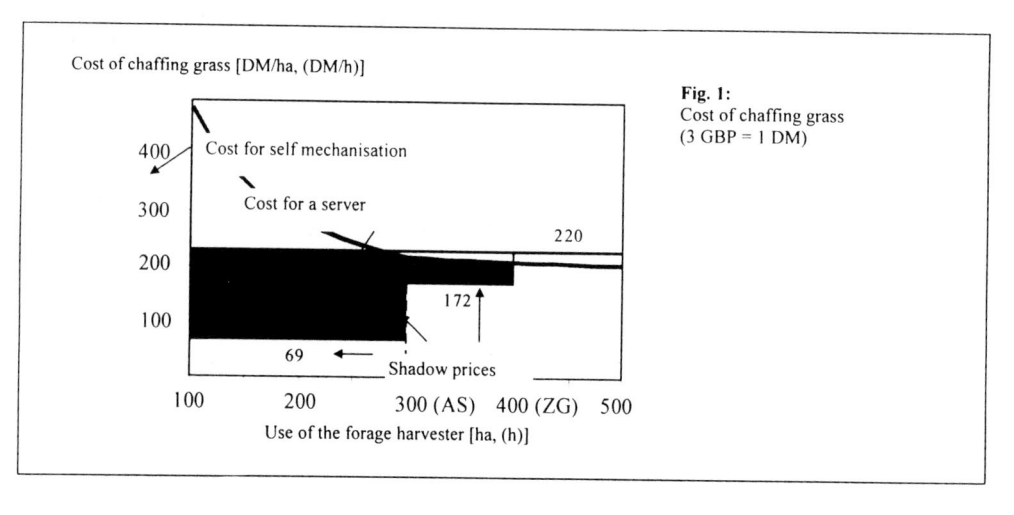

Fig. 1:
Cost of chaffing grass
(3 GBP = 1 DM)

The Interactive Footpath: an online tool for the land based professional

Felton, T., Stone, M.

Seale-Hayne Faculty of Agriculture, Food and Land Use, University of Plymouth, Newton Abbot, Devon. TQ12 6NQ. United Kingdom.

The student centred learning initiative at the University of Plymouth has identified the following characteristics as core to the concept of student centred learning.

Namely that Student Centred Learning is an active and dynamic process through which students develop deep approaches as learners, taking responsibility for their own learning.

The aim of the Initiative is the further development of Student Centred Learning to provide students with the best possible educational experience in a flexible, stimulating environment which enables students increasingly to:

- have access to the resources that will help them most as individual learners;
- interact with staff and collaborate with other learners;
- develop skills and self-awareness of their own learning process;
- develop increasing independence in their learning;
- reach their potential in the subjects studied.

This paper examines the way in which information technology, in particular the use of the internet, can be used to enhance a traditional case study approach to investigating the law relating to public access to the countryside. It demonstrates how a greater degree of flexibility may be offered to students in their approach to learning. The paper also examines the concept of lifelong learning in the context of an educational administrative system that puts rigorous time limits and deadlines on taught modules. The paper demonstrates a way in which technology can be used to help break down the concept of the module as a discreet and contained experience in order to assist the student on their path towards lifelong learning.

In order to meet customer demand, serve the needs of prospective employers and the communities which the University serves, the validation and introduction of new modules becomes an imperative. This paper analyses the approach adopted by a subject team when designing a new module in Environmental Law for delivery to students on the BSc in Rural Estate Management. The paper reflects on how the module design and implementation satisfied the teaching strategies and philosophy of the University of Plymouth and the requirements of the Royal Institution of Chartered Surveyors (RICS), the professional body that accredits the Rural Estate Management degree. Further examination is made of the lessons learned in this experience and the student feedback to the first running of this module which required the traditional skills of oral debate and report writing alongside the use of video conferencing.

This paper is one of a series that will be delivered using a Student Centred Approach.

The Interactive Footpath: an online tool for land based professionals ~ Felton, T & Stone, M University of Plymouth, UK

Footpath Development

The Interactive footpath has been in existence for over three years. It has emerged from the teaching, learning and research interests of the development team.

The footpath seeks to make full use of Internet technologies, to provide and facilitate an exchange of information between the many individuals interested in this area of rural management and law. It was originally created with a funding grant from the University Continuing Vocational Education fund.

Aims & target audience

The Footpath seeks to provide:
- An interactive learning approach
- A research and communication facility
- An identifiable and coherent Web Site with regard to countryside access in England and Wales

The footpath was created for three key groups:
- Students and Academics
- Legal Professionals
- Land Based Professionals

Although there is no substitute for getting your boots muddy, the Interactive Footpath does gives the above groups access to Rights of Way issues in context and on a continuously developing basis. For example users can see and read the issues relating to broken stiles, blocked paths and vehicular use of bridle ways regardless of the time of year, state of the weather or distance from accessible country.

What is on the web site

You will be able to take a leisurely walk through the Devon countryside which will allow you to discover:
- Legal principles relating to the public rights of way
- Techniques for maintaining those rights of way
- Useful links to web sites related to: Countryside Access • Access Policy • Legal Developments • Partnership Initiatives • User Groups • Latest Initiatives • Maintenance
- Academic Articles

Training Package

The Footpath contains an introductory learning package, the Footpath Quiz, to allow users to extend their knowledge of public rights of way and questions relating to countryside access by making full use of the Interactive Footpath web site.

Questions relate to the following topic areas:
- The Definitive Map
- Landowners Obligations
- Highway Authority Obligations
- User Rights and Obligations
- Maintenance of Public Rights of Way
- Obstructions of Rights of Way
- Waymarking
- Modification and Diversion Orders
- Voluntary Access Schemes
- Pressure Groups and Policy

The questions relate to the web site and highlights the various areas that need to be considered when undertaking a rights of way survey.

The site also provides useful links and literature sources relating to the site as a whole and the training package.

Working online

The footpath can be found at:
www.pill.plym.ac.uk/footpath

The Interactive Footpath is hosted by PILL, The Plymouth Internet Learning Lab.

5th European Conference on Higher Agricultural Education

The development and use of a managed electronic learning environment for agricultural education and training

Stone M., Witt, N.

Seale-Hayne, Faculty of Agriculture, Food and Land Use, University of Plymouth, Newton Abbot, Devon. TQ12 6NQ. United Kingdom.

Advances in communication and learning technology are allowing easy access to the Internet from anywhere in the world. This has fuelled interest, demand and research into methods of using and enhancing the technology to deliver a range of education, training and continuing professional development (CPD). The pressures of time and resources to allow personnel to attend programmes of education and training has led to a research programme being instigated at the University of Plymouth to investigate cost effective methods of delivering material to both on campus and remote learners.

Technology now allows remote learners to interact with both their tutors and their peers whilst accessing a variety of learning resources including text, pictures, graphics and video. Recent research has shown that the user requires more than just access to courseware and the associated material. This paper will explain the approaches taken to develop a suitable operating environment and interface for electronic distance learners. This environment whilst containing the course material offers a support structure, feedback mechanisms, essential resources and administrative protocols. The authors' research and development of the necessary systems have involved:

- investigation into the feasibility of an integrated delivery system;
- the piloting and delivery of online learning;
- furthering links with external bodies for delivering tailored CPD;
- the construction of a professional and easily navigated working environment.

Key outcomes include:
- a suite of modules for education, training and CPD;
- guidelines and best practice for delivering material at a distance by electronic means;
- models and paradigms which allow new material to be inserted into a proven framework

Through this paper the authors will present the rationale behind the system and processes involved, share knowledge required in designing such an integrated system and present what they believe to be an approach with widespread applicability to the design and delivery of teaching and training.

The development & use of managed electronic learning environment for agricultural education and training

Stone, M and Witt, N, University of Plymouth, UK

History & development

The authors have been involved in the design and delivery of electronically supported learning since 1995. This started with the creation and testing of a suite of modules for education, training and CPD (continuing professional development). What was rapidly apparent was that however good the learning materials, learners and tutors required a suite systems and structures to give them the confidence and support they needed to make the the most of the online learning experience.

This work led the the creation of the PILL managed electronic learning environment - The Plymouth Internee Learning Lab: pill.plym.ac.uk

Integrated features of PILL

- ❖ A professional, easily navigated working environment
- ❖ Cost effective methods of delivering education, training & CPD to both on campus a remote learners
- ❖ Learning resources including text, graphics and video.
- ❖ Learner interaction with both tutors and peers
- ❖ A support structure for electronic distance learners
- ❖ Assessment and feedback mechanisms
- ❖ Learner tracking
- ❖ Essential resources and administbative protocols
- ❖ Guidelines and best practice for delivering material at a distance by electronic means

e-course management steps

- ❖ Defining and publishing policy and procedure
- ❖ Remote management & administration
- ❖ Asynchronous administration of asynchronous learning patterns
- ❖ The roles and responsibilities of support services

e-learning policy

- ❖ Clear policies are needed to integrate e-learning with institutional services:
    - Registration systems
    - Library service provision
    - Computer networks
    - Course payment systems
    - Learner records systems

e-learner support

- ❖ Learning to learn online
- ❖ Clarity of support & speed of response
- ❖ Asynchronous support
- ❖ Peer support facilitated
- ❖ Library and resource provision

e-group work

- ❖ Staff development in networking tools and techniques
- ❖ Networking of development, support & delivery teams
- ❖ Guide and support learners e-group working through course design and administration

e-learning materials

- ❖ Design the e-learning process first
- ❖ A clear assessment strategy is vital
- ❖ Then convert & adapt existing material
- ❖ Embed course material in a programme of study
- ❖ Creation materials to fit a house style
- ❖ Instructional Management System

e-assessment & tracking

- ❖ Assessment strategy must come early in course design
- ❖ Assessments need to be frequent with quick feedback
- ❖ Some formative assessment may be automated
- ❖ Facilitation of group work must come early in a course
- ❖ Tracking of registration, progress, assessment and communication supports:
    - Learner information systems
    - Course management
    - Course performance feedback to learner, tutor and administrators
    - Quality and external examination systems

5th European Conference on Higher Agricultural Education

PILL

Overcoming Marginality: Cyber-Training in Cider Country?

David, M.

University of Plymouth, Plymouth, Devon. PL4 8AA. UK.

This paper seeks to outline findings from a number of research projects looking at the capacity of Information and Communication Technologies (ICTs) to assist in rural training development. The paper will initially examine the results of the author's own research into the Rural Area Training Initiative (RATIO) and the ADAPTthroughRATIO programme that followed from it. In establishing a series of 'telecentres' and distance learning packages for use across the South West of England (Devon, Cornwall and West Somerset) these twin projects enabled research to be conducted into the capacity of ICTs as vehicles for distance learning, and, potentially, as a means of overcoming the training gap experienced in areas of rural marginality. This research sought to examine the capacity of electronic networks to facilitate 'learning networks', as well as the role of 'tacit knowledge' in building and inhibiting distance learning. In particular, this research focused upon the relationships between divergent communities: those of product developers, centre managers and centre users. The abstract below outlines the findings of this research. The paper then addresses other research in the area. Links between the author's research and other work in the field will be highlighted, general conclusions drawn, and suggestions for further research in the field put forward.

The author's own research ran in two stages (for part one see David *forthcoming*, and for part two see David 1999). The first part of the research involved sampling a RATIO centre in each of the three counties and conducting interviews with local people (specifically employers and employees from Small and Medium Sized Enterprises). These interviewees were not users of the RATIO centres as such though some had used centre services and most had heard of the centre in their area. The aim of this research was to evaluate the social and economic conditions into which the centres were hoping to introduce their services. It was also an attempt to examine what local 'potential' users thought they needed and what they believed they could afford, especially in terms of time and effort. The second part of the research focused upon users of centre services and the sample of centres was expanded to include three centres from each of the three counties. What did users make of the format of distance learning, what had they expected, what did they want, did they get it and how did they think it could it be done better in the future? There were two questions that guided the research: 1. Just how far can the abolition of space and time constraints promised by new ICT aid in redressing economic and geographical marginality? 2. Just how far are the weaknesses produced by economic and geographical isolation in fact reproduced within virtual domains?

In a simplistic fashion one might suggest that the function of AtR must be to bring expert knowledge within the reach of local users. How well users, centre managers and product developers worked together was at the centre of the research. Users are not stupid. There is a tendency to assume that users (and potential users) are ignorant of their own best interests. This deficiency model is inadequate. There are other limiting factors at work besides limited information. In addition to a lack of knowledge there is also a knowledge of lack. SMEs in marginal areas experience difficulties devoting time to training where low levels of staffing limit free-time, and where the nature of part-time and seasonal labour markets make training staff problematic. Finally, many businesses in the craft, agriculture, tourism and leisure sector do not want to be 'more efficient' by choice, not out of ignorance. I will expand on this in the paper. Product developers have to recognise the local knowledge held by users and potential users. To dismiss local perceptions as simply the product of a lack of knowledge ignores limitations of other (structural) kinds and forms of orientation and culture that deviate from classical economic models of rational action. It is also important to orient towards the expertise that exists within local communities. If products are to be adapted to local needs then the role of 'bare-foot tutors' (of which I will say more) is crucial. RATIO centres play a crucial role as co-ordinators, facilitators and promoters, building the networks of local expertise, assisting beginners overcome their initial 'fear of flying', and reaching out to those who have yet to participate in terms that those potential users will find meaningful and relevant. Centres play a crucial mediating role between a range of groups, developers, users, local experts and potential users, whose fears and needs may otherwise go unheard. The role of centres should be as sites for the transmission of tacit knowledge. This is a far cry from a 'cash-machine' model of local e-training provision. While no one would suggest such a model out loud the limited funding available to local RATIO centres tends to such an orientation by default. As such, while accidental, the outcome of funding allocation and competing interests, was to severely limit the effectiveness of the programme. Developers, as designated holders of expert knowledge, need both to improve their means of listening to local users and centres, and their methods of communicating to local users, potential users and centres. Designated expertise carries with it the danger of insularity. The pressure towards a market orientation in education increases the pressure on product developers to orient towards those that can afford to pay, so reinforcing the gap between developers and those that AtR sought to serve.

The limitations to be found in the application of distance learning are very real. However, contrary to the beliefs of techno-sceptics, such limitations are not total and/or intrinsic. An understanding of the knowledge and circumstances of local users and potential users, the knowledge and circumstances of developers and those of prospective intermediaries, enables a better understanding of both the potential and failings of specific distance learning projects.

Whilst the author's own research was not specifically focused upon the agricultural sector the recent work of Warren (et al 1996, 2000) highlights many parallels with the findings of the author. Exploring these parallels will form the second stage of this paper. The third stage will seek to examine how the study of these parallels may improve our understanding of the tension, raised above, between a 'cultural deficit model' of weak ICT take up within the rural economy (lack of knowledge), and a 'material deficit model' (knowledge of lack). I suggest that this tension, and the failure to address it effectively, is the most significant factor in explaining the limited success of interventions that seek to overcome rural marginality by telematic means.

Bibliography

David, M., 2000 (forthcoming). Lost in Cider-Space. Systemica.

David, M., 1999. Cyberculture in Cider-Country. Adapt, Survive & Thrive: Proceedings of the 1999 ADAPTthroughRATIO Dissemination Conference, Plymouth, August 31st – 1st September 1999, pp. 135-145.

Warren, M., 2000. Evaluation of the South West Agricultural and Rural Development Project (Stage 1 Report). University of Plymouth.

Warren, M et al., 1996. The Uptake of new communication technologies in farm management. Farm Management, 9, 7, pp. 345-356.

The Theoretical Underpinnings of Student Centred Learning.
A case study based on business management teaching and learning for agriculture and related programmes of study at the Seale-Hayne Faculty, University of Plymouth.

Stone, M., Williams, R., Soffe R., Fisher, S.

Seale-Hayne Faculty of Agriculture, Food & Land Use, University of Plymouth, Newton Abbot, Devon. TQ12 6NQ. UK.

In the 1930s Seale-Hayne was at the forefront of Student Centred Teaching with programmes designed to equip students for the real world.

Even then we were using new technology (The Farmers Boy film) to promote the understanding of and encourage the uptake of agricultural education.

In the last five years the emphasis has changed from Student Centred Teaching to Student Centred Learning. Student feedback suggests Student Centred Learning works well at the module level. However, the challenge is to build Student Centred Learning into pathways and programmes in such a way as to assist student motivation and development.

This paper will focus on the priorities of the Student Centred Learning initiative / especially for business management teaching.

This paper will also address how this Faculty and institution-wide initiative sought to:
- help staff focus on and engage with a strategic process
- involve all staff
- build on existing teaching and learning good practice
- make Student Centred Learning a transparent and coherent experience for students
- implement Student Centred Learning for whole programmes of study.

Key methods discussed include:
- the creation of a Faculty Student Centred Learning Team
- the creation of a Faculty Teaching & Learning Group
- the role and support of teaching teams and Pathway Coordinators
- the creation of prioritised academic and resource plans
- the use of web-based support
- the use of Student Centred Learning exhibitions.

This paper is one of a series that will be delivered using a Student Centred Approach.

Indicative Reference
Teaching Methods in HE: a glimpse of the future by Bourner & Flowers 1997 in Reflections on HE.

The theoretical underpinnings of student centred learning

Stone, M, Williams, R, Soffe, R, and Fisher, S, University of Plymouth, UK

What do we mean by SCL

SCL is about getting students to...
- think for themselves
- recognise learning opportunities
- learn flexibly
- learn independently
- work together
- relate to the real world

SCL over time

SCL at Seale-Hayne was not and is not new. A student centred approach to learning is very much part of our institutional heritage and it remains deeply embedded in our organisational culture.

In the 1930s Seale-Hayne was at the forefront of Student Centred Teaching with programmes designed to equip students for the real world. Even then the Faculty was using new technology in the form of a short film, 'The Farmers Boy', to both promote and make our programmes more accessible.

However in the last five years the emphasis has changed from Student Centred Teaching to Student Centred Learning. Student feedback had suggested that Student Centred Learning was working well on an add hoc individual basis at the module level. The challenge was to build Student Centred Learning into pathways and programmes in such a way as to assist student motivation and development.

A new approach to SCL

What was new, was the aspiration that:
- SCL should be a transparent and coherent experience for students

- SCL should be designed and implemented for whole programmes of study

- SCL was to be University wide funded initiative

Working strategically

There were a range of organisational actions designed to launch the initiative:
- Creation of Faculty SCL Teams
- Creation of Faculty Teaching & Learning Groups
- Creation of prioritised academic and resource plans

Within the Faculty specifically the SCL initiative was driven by:
Pathway Coordinators and their teaching teams
Supported by staff paid for the SCL initiative
The use of web based learning support materials and systems
The use of Student Centred Learning exhibitions

Transparent and coherent

The key priorities of the Student Centred Learning initiative for the business management teaching team were to:
- Standardise policies, procedures and methods of learning design and delivery based on shared best practice

- Initiate a strategic pathway wide process for the design and operation of all management modules

Key Outcomes

The key outcomes for the business management teaching team have been:
- I much higher level of effective team working
- A wealth of new ideas
- Greater engagement of students with the subject matter
- Reduced staff and student frustration and confusion

5th European Conference on Higher Agricultural Education

BMGT

Student Centred Learning in Practice.
A case study based on business management teaching and learning for agriculture and related programmes of study at the Seale-Hayne Faculty, University of Plymouth.

Stone, M., Williams, R., Soffe, R., Fisher, S.

Seale-Hayne Faculty of Agriculture, Food and Land Use, University of Plymouth, Newton Abbot, Devon. TQ12 6NQ. United Kingdom.

This paper will outline examples of Student Centred Learning (SCL) in practice from all three years of the taught undergraduate programme at Seale-Hayne Faculty, University of Plymouth.

SCL is an approach to designing and delivering learning opportunities. To introduce Student Centred Learning in practice it is illustrative to answer the question 'Can reading be Student Centred?'

Simply telling students to read a specific text is not Student Centred. However it becomes Student Centred Learning when the reading is accompanied by further information or is related to a task, for example:
- how the reading material relates to the module
- an activity to do related to the reading
- a question to answer while reading the material
- alternative sources or media covering the same area, this helps deal with difficulties of material availability and style.

This guidance should be targeted to make sure that the students have both motivation and focus in their work.

Many staff and students think of SCL on a module basis. As the example above suggests SCL can operate on an individual activity scale.

While any learning activity can be designed to be Student Centred this paper will focus on whole module approaches and their integration within an overall pathway approach to the design and delivery of Student Centred Learning.

The paper will address the key issue of consistency and coherence within curriculum design and go on to show how the key outcomes of Student Centred Learning are planned for, namely:
- getting students to think for themselves
- getting students to recognise learning opportunities
- getting students to learn flexibly
- getting students to learn independently
- getting students to work together
- getting students to relate to the real world

This paper is one of a series that will be delivered using a Student Centred Approach.

Indicative Reference
Teaching Methods in HE: a glimpse of the future by Bourner & Flowers 1997 in Reflections on HE.

Student centred learning in practice
Stone, M, Williams, R, Soffe, R, and Fisher, S, University of Plymouth, UK

SCL learning programmes

Detailed below are three examples of Student Centred Learning in action. Each of the examples involves the use of technology.

The aim of this work has been to:
- enhance the on-campus student experience
- extend access to higher education
- support professional updating and training

BMGT online

BMGT online is an information and resource bank developed by the Business Management Pathway team. It is designed to be a reference point for both students and staff. The aim is to improve the student learning experience through the delivery of a more clearly coherent pathway.

The aim was to achieve a single, simple set of instructions to which all BMGT modules can subscribe. This development reduced complexity and provided increased consistency within and between BMGT modules.

Take up was facilitated by integration with computing induction exercises.

BMGT online includes information and resources on:
* Module information
* Study and employment skills
* Teaching methods and ethos

MSc Marketing on CD

The MSc in Marketing at Seale-Hayne is a programme giving students the opportunity to study at a distance using CD and web technology.

Students are assisted in their route through the programme by a video guide on the CD.

The course materials are based on real businesses and real business situations.

The Internet for business

Using the Internet for Business, an online and face to face programme delivered to over a thousand small businesses

The Objectives is for users to gain an understanding of how the more popular Internet applications work and at the same time get an insight into how these applications can be applied in the business world in a productive way. Users completing the course were able to enter the world of e-commerce armed with this knowledge and thereby make their business dealings on the Internet more productive.

This course has been tested on several small groups of students in a classroom based situation. It has also been externally evaluated. The course does not have any sort of academic award scheme, although it has been designed to be equivalent to NVQ level 3/4.

The course includes a number of assessments which must be completed before further course materials can be accessed. The assessments are recorded. Course participants need no knowledge of the Internet to get started - and some parts are aimed at absolute beginners. However, some knowledge of computers is assumed and some of the technical courses might be quite difficult for those with limited computing skills.

Online learning androgogy

Work is also being undertaken at the University by the Communication & Learning Technology Research Group (CoLT). www.colt.org.uk which investigates the undertaking of tailored CPD delivered online.

5th European Conference on Higher Agricultural Education

Developing Student Centred Learning in the Information Age. A case study based on business management teaching and learning for agriculture and related programmes of study at the Seale-Hayne Faculty, University of Plymouth.

Stone, M., Williams, R., Soffe, R., Fisher, S.

Seale-Hayne Faculty of Agriculture, Food and Land Use, Plymouth University, Newton Abbot, Devon. TQ12 6NQ. United Kingdom.

This paper will outline a development programme for Student Centred Learning. This will include an exploration of how learning technology can and is being used to:
- enhance the on-campus student experience
- extend access to higher education
- support professional updating and training

While any learning activity can be designed to be student Centred this paper will focus on whole module approaches and their integration within a pathway approach to the design and delivery of Student Centred Learning.

The key examples examined by this paper are:
- BMGT online, an electronic learning support tool
- MSc Marketing, a programme giving students the opportunity to study at a distance using CD and web technology
- Using the Internet for Business, an online and face to face programme delivered to over a thousand small businesses

This paper is one of a series that will be delivered using a Student Centred Approach.

Indicative Reference
Teaching Methods in HE: a glimpse of the future by Bourner & Flowers 1997 in Reflections on HE.

Developing student centred learning in the information age

Stone, M, Williams, R, Soffe, R, and Fisher, S, University of Plymouth, UK

Introduction

The use of technology in education is becoming more pervasive. Key ways in which technology is used to support learning include the breaking down of time, distance and location access barriers.

Communication and Learning Technology will...
- change the way students approach their studies
- change the way staff deliver and manage learning
- change the organisational and operational culture of universities

Outlined below are some of the key issues involved with the design, delivery and management of learning through the medium of technology. Much of this work is distilled from the experience of the Communication & Learning Technology Research Group (CoLT). www.colt.org.uk

e-course design requires

- Application of educational good practice
- Moving from a face to face teaching norm
- Project management of the process
- Dealing with the fact that few staff will have all the skills
- Dealing with whole courses/programmes of study
- Clear links with support services
- Use pervasive technology & software

e-course creation requires

- Creator motivation and experience
- Treating developers as learners
- Piloting and getting feedback
- Scalability and sustainability
- Coherence & consistency of materials and modules
- Clear conversion and adaptation strategies

e-course resources issues

- Ownership - working with 3rd parties
- Updating
- Access formats
- Collation, navigation and links
- Library material and service provision
- Copyright
- Student generated resources

e-learning the market

- A quality product is not enough to ensure take up
- Market research is the weakest link in most e-learning developments
- Many developments are supply side driven
- e-learning is a solution to a problem that the target audience may not know it has or be ready to tackle

e-learning marketing

- Confidence is required in the course, the provider, the technology and themselves
- Promotion needs a working model
- Purchase decisions are based on more than just the course / materials
- The purchaser may not be the learner
- Provider & learner identity is crucial to the sale

e-learner needs

- Confidence in the course
- Confidence in the course provider
- Confidence in their ability to learn at a distance
- Confidence in their ability to use the technology
- Personal & organisational spin offs
- An organisational context for their work
- For learning material to have a context
- For learning to be a social process
- Peer support / learning

e-learner demands

- Clarity of what is on offer
- Clear benefits to the learner
- Quick registration (15mins max)
- Clarity of navigation & progression
- Quick response to questions
- Understanding of learner technical problems
- High presentation standards

5th European Conference on Higher Agricultural Education

Student Attitudes to Student Centred Learning.
A case study based on business management teaching and learning for agriculture and related programmes of study at the Seale-Hayne Faculty, University of Plymouth.

Stone, M., Williams, R., Soffe, R., Fisher, S.

Seale-Hayne Faculty of Agriculture, Food and Land Use, University of Plymouth, Newton Abbot, Devon. TQ12 6NQ. United Kingdom.

Getting students to focus on their learning and provide meaningful feedback upon their learning experience is a difficult process. This paper addresses the approaches taken and the results generated from the Student Centred Learning initiative at the Seale-Hayne Faculty.

Key approaches that will be addressed include student focus groups dealing with the questions:
- What activities help you to learn?
- What has been your best learning experience?
- What can help effective learning?
- What will be the importance of IT in teaching in the future?

The paper will also deal with student reactions to:
- the use of technology
- distance learning.

Much student feedback relating to Student Centred Learning is positive, especially at the end of the process. The challenge faced in this case study has been to move students and staff from a learning comfort zone into a more effective learning culture.

The paper will conclude by looking at the results of efforts to:
- be proactive in the gathering of feedback
- 'front load' teaching with information and guidance on the learning process
- encourage more staff student interaction
- build student discussion and understanding of the learning process into both induction and specific modules.

This paper is one of a series that will be delivered using a Student Centred Approach.

Indicative Reference
Teaching Methods in HE: a glimpse of the future by Bourner & Flowers 1997 in Reflections on HE.

Student attitudes to student centred Learning

Stone, M. Williams, R. Soffe, R. and Fisher, S. University of Plymouth, UK

Methodology

- To explore student attitude to learning
- To get students perceptions of the learning methods used at Seale-Hayne

Sample
1 in 7 invitations to the 829 BSc students not on industrial placement

2 pilots were conducted with students from different stages. The main focus group consisted of one first year group, one second and two final year groups. The poster includes student quotes from all the focus groups. All groups were facilitated by postgraduate students.

The focus groups covered a wide area of learning techniques. For those interested in finding out more the transcripts can be viewed on N6. A booklet containing more student quotes on SCL issues only, can be found on the table beneath the poster.

What activities help you to learn?

Lectures

Some lecturers have got a schedule to be done and they hang on to it for grim death and everyone is falling asleep like several hours ago, and they are determined that they are going to finish what they want to say.

Make it interesting by relating the industry rather than trying to explain something and you can't relate it to anything. They could say 'while I was doing this and it makes sense to do it this way'.

I use lectures... as the basis then go away ... and use books recommended or the Internet.

I think it would be much more useful to have all the OHPs in a handout and work through the handout so you're reading it as he's explaining it and then you have space between each question to write down the notes that you feel beneficial from what the lecturers saying under the basic handout rather than just bombarding us with OHPs.

Seminars

You need to know how to do seminars and talk to people.

I think they're vital for the future. Your always going to have to do something with an interview. A lot of people ask for a presentation and if you haven't had past experience for that your in trouble.

Also with seminars you're vocalising things often in a group, so to discuss something with the group makes you bounce ideas off each other and come at it from a different angle and it's got something that seminars haven't and vice versa. It opens the subject.

Group work

Complicated tasks can be achieved more easily - goals are achieved more easily because work is shared

More group work in the first year would be useful because it teaches you to work in a team.

It's good practice to be thrown with a group of people that you don't know because that's how the real world works. You have to get on with people.

It's another life skill.

Practicals

It would be good to get the industry involved with it. Perhaps the industry could set the assignment based around what they want to achieve so you can make it more relevant

I think it's really helpful to talk to the people who are actually out there doing it because sometimes I've sat in a lecture and listening to the lecturer and it's making sense and everything but I'm thinking 'if you know all this why aren't you out there making loads of money'?

Talking with colleagues

I find that I end up learning more from talking to one or two other people on how to do the assignments, if the lecturer hasn't been very clear. We do a little brainstorming session with two or three people.

Best learning experience?

Being on placement because I got a lot of practical knowledge

Probably I think Honours Project. That's quite an involved learning experience

... is a horrible subject to teach, it's boring and it's so diverse, and he tries to make it funny with examples - funny stick men and clips - to keep people interested

I think Bosscat is brilliant.

Evaluation or farm visit or forestry practical, anything like that I can remember. I think, just about everything we've covered on those whereas things actually in lectures being bla bla bla'd out all the time.

So I think anything that's hands on and practical is definitely the way forward.

Group work has been the best because everybody in a group knows something that somebody else doesn't know so you learn from each other

What can help effective learning?

Some of our best lecturers are the part time lecturers that actually do the job. There up to date and they also an amazing amount of views which some of our lecturers haven't got.

I think it's pretty much all down to self motivation.

I think you learn more by discussing things and then thinking about it afterwards.

Debriefs are excellent.

He just went through the class like we are now and we were all bouncing off each other and I can honestly say that I remember most of what he said.

Importance of IT in teaching in the future?

We're all going to need it in the future. It's something that's going to continue to expand

If you could go to the Internet and you could study the lecture through notes off the Internet in my mind that would make it a lot easier for me because I could work through at my own pace.

Facilitators comments

The focus group provided thoughtful discussions amongst the students

The attitude of the students was a lot more positive than I thought it would be

The students were very forthcoming with their views on their learning experience at S-H and were keen to make constructive suggestions. Some more practical than others, on how to improve that experience

My group, 1st years, were just coming to grips with the learning process & having to adjust to life at S-H

The Focus Groups provide the opportunity for students to express their opinion in a less inhibited way than within more formal channels, with the benefit of anonymous feedback to policy makers and strategies

I think one of the most predominant findings in my groups was the importance that they gave to field work. More of it please. The third year's were a very positive and committed group.

5th European Conference on Higher Agricultural Education

Posters

Adapting the curriculum of higher agricultural education in transition

Kartashevich A.N.

Belarusian Agricultural Academy, Lenin Lane 4, Gorki 213410, Belarus.

The right choice of teaching technologies involving educational, information and technical aids in teaching, the process of teaching, means of knowledge control, duration of particular types of teaching process is one of the important factors in reforming the system of higher education started in Belarus and other countries of CIS.

From the point of view of a systematic approach the teaching technology can be considered as hierarchical multilevel structure being studied from different sides by different sciences, but the nature of relationships being studied by a particular science greatly depends on the degree of its development and research methods used. For all that the complex object is considered as an interacting complex of elements subjected to the supplementary impact of external factors and having parameters of initial and final status.

This complex of factors requires consideration of the following :
- applicants' level of knowledge
- the final level of knowledge
- choosing a teaching technology:
    - resources and staff training. - optimal resources,
    - optimal results,
    - the degree of specialisation and its timing,
- levels in the curriculum:
    - strategic,
    - tactical,
    - working.
- the method of the curriculum:
    - fixed,
    - flexible.
- the method of teaching
- the structure of the curriculum:
    - composition of disciplines,
    - number of disciplines.
- assessment:
    - final or continuous?
    - types of knowledge assessment.

This paper explores these factors in the context of rapid changes which are occurring in higher agricultural education in Belarussia.

Qualitative Model of Assessment of Results of Higher Agricultural Education

Kasimov RY[1], Gotovtseva I P[2], Akhmetov R G[3]

Department of Methodical-Educational Section of Moscow Agricultural Academy, Russia[1]
Department of Pedagogical Faculty of Moscow Agricultural Academy, Russia[2]
Pro-rector Assistant in Economic Problems, Moscow Agricultural Academy, Russia.[3]

Modern methods and means of cybernetics enable the active treatment of results of mass measurements. Pedagogical-qualitative information is about the quality of higher educational institution activity and is also ascribed to quantitative information. In this presentation we examine such aspects of higher education institution activity as the formation of a personality of a specialist in agriculture whose inner motives will harmonise his/her vital activity with nature and society and will not let it contradict with human life. Results of many years of research have revealed that one of the most serious disadvantages of using cybernetics means and methods for management of quality of education and formation of personality of a specialist within the precincts of the university is not efficient modelling of the quality of education altogether. In this report produced from the philosophical outlook of completeness, non-discrepancy and independence of ecological, economical and physiological indices we tried to formulate indices of effectiveness of higher agricultural education. The model is constructed with the help of the theory of multitude and Large Numbers' Law is the principle methodological basis of it.

In this connection the task to improve the effectiveness of management of higher agricultural education with the help of cybernetic means and methods is considered to be very urgent. This task can be formulated as philosophical - logical substantiated construction of a model of qualitative indices of effectiveness of higher agricultural education [1]. In this paper some conceptual approaches to construction of these models are described [2]. This modelling is based on the unity of ecological, economical and psychological indices of results (effectiveness) of higher agricultural education (world outlook aspect). At the same time the modelling is based on the well-known philosophical principles such as completeness, non discrepancy and independence. In this case Ailer's diagrams are considered to be an adequate modelling instrument. It is worth while reminding that the excellent presentation of Mr. Charles Maguire, the representative of World Bank (USA) at the plenary session of the 4th European Conference on Higher Agricultural Education (Moscow, 1998) was a fine illustration of the correct application of Ailer's diagrams for modelling of assessment of results of financial support of agriculture.

References

(1) Kasimov R.J. The principle of the unity of intuition of the skilled teacher and the algorithm of measurement in pedagogical qualimetry. - Moscow Agricultural Academy alter K.A.Timiryasev, 1999, 151 p., Literature cited: 337.

(2) Kasimov R.J. Methodological basis for construction of automated systems of management of students' study on the basis of personal computer. - Moscow, Publishing House "Petite", m., 2OOO 156 p.

The agrarian production ecologization in the Baikal region as the basis for transition to sustainable development.

Korsunova T.M., Popov A.P., Tumanova M.B.

Buryat State Agricultural Academy, Ulan-Ude 670024, Russia.

The Baikal Region comprises the world famous Lake Baikal and a reservoir declared by UNESCO the Region of the World Natural Heritage. The region is of particular importance and great value for the planet's biosphere structure. The unique natural objects, enormous resource potential, high endemic biota, regional peculiar synthesis of cultural wealth of Buddhism, Christianity, ethno-cultural and material morals and manners predetermined the Baikal Region to be chosen as a world model area for sustainable development, a peculiar ecological civilisation micromodel. The sustainable development conception is based on balanced ecological and economic interests, ecologization in all spheres of human activities, including agriculture.

Agrarian sphere being of vital importance in the sphere of anthropogenic activity has multi-planned and often negative impact on natural ecosystem. Man and nature discord has the salient revealing exactly in the agricultural sphere. Industrialising of modern agricultural production with applying power-consuming and environment-destroying technologies, fertilisers and pesticides overdosing, wide-scale land reclamation, ignoring of regional economic peculiarities results in upsetting natural ecosystem balance, sharp decreasing of agro-ecosystem ability to maintain ecological balance by self-regulation means.

The given ecological problems in connection with the agrarian nature utilisation are quite characteristic of the Baikal Region and increasingly due to the region's natural and climatic peculiarities are revealed extremely sharp. The territory of the Baikal reservoir belongs to "risky farming" region: sloping nature of the locality and small amount of precipitation (200-400 mm per year), mostly of heavy shower character, shallow snow covering and withered spring winds, shallow granule-metric soil structure with little humus, poor natural vegetation land surface. All these factors caused natural ecosystem fragility and vulnerability and mostly its edaphic component-soil. The land tenure ecological crisis is more evidently seen in topsoil, which is easily subjected to degradation with the slightest violation of soil-protective farming technologies. Primarily, it resulted from the regional socio-economic conditions of farming development. Changes of the traditional farm running, with plant-growing being of primary importance, resulted in increasing arable areas (by 400,000 ha.) after having been ploughed up virgin sandy soils with unstable erosion. The increasing the total number of sheep with poor forage reserves and excessive pasture loading, violating farming agro-technologies when ploughing steep sloping lands resulted in increasing erosion and degradation on ploughs and pastures.

Thus, farming running with violating nature-protective requirements resulted in revealing intensive erosion and degradation processes in the Baikal Region, upsetting the ecological balance, visible landscape reconstructing and in many places - even absolute agro-landscape degradation (on ploughs and pastures) and evident negative impact on the unique lake ecosystem. In order to preserve the World Heritage Region it's really necessary to develop the strategy for ecologically sustainable agriculture based on the total ecologisaton of agricultural products manufacturing. Farming organisation, sowing, cattle-breeding objects, applied machines and technologies should be environmentally-friendly. Special attention should be devoted to restoring traditional farming systems to imitate care of soil and to preserve sustainable ecosystem. Ecologisation aims at creating new, up-to-date and prospective farm machines to evade machine soil degradation (dispersion, compression, erosion), creating special machines for sloping lands prevailing in the region. The significant direction is considered to be rational utilising of production wastes, applying of resource-saving technologies.

Therefore, sustainable agriculture should be founded on resource saving, maintaining agro-ecosystem productivity on the basis of balanced ecological, social and economic factors to provide long-term sustainability.

International Hydrology Course – an example of a water-orientated postgraduate study programme at CUA Prague

Kovar, P.

Czech University of Agriculture in Prague, 165 21 Prague 6 – Suchdol, Czech Republic

In the framework of harmonisation of agricultural and forestry production with land and water management, the Forestry Faculty of the CUA Prague organises a water-orientated postgraduate Courses. These training courses titled „Hydrological Data for Water Resources Planning" (held in English) are organised every even-numbered year under sponsorship of UNESCO and WMO. There are designed for those specialists in hydrology who desire their knowledge on monitoring, processing and use of hydrological data for water resources planning and management preferably in agriculture and forestry. The Course duration is two months: the last session has been organised from May 17 to July 12, 2000 with 20 participants from 11 countries. Lecturers from three Czech Universities (CUA Prague, Charles University and Czech Technical University) delivered lectures, seminars and practical training to participants on the subjects below.

The Course is structured in four parts

A. Data and Computing
B   Hydrological Processes and Modelling
C   Water Resources Management and Real Time Forecasting
D   Case Studies, Hands-on Workshops and Projects

List of subjects

A
1 Statistics and stochastic processes
2 Meteorology, climatology and global change
3 Hydrological data collection and GIS

B
4 Engineering hydrology and hydraulics
5 Hydrogeology and subsurface flow
6 Water chemistry and quality
7 Hydrological models

C
8 Reservoir operation water resources management
9 Hydrological forecasting (floods)
10 Environmental impacts

D
11 Use of PC and internet
12 Hydroinformatics
13 Water resources management in agriculture
14 Case studies, hands-on exercises
15 Guest lectures

The address of the course is: Department of Water Resources,
Department of Land Use and Improvement
Faculty of Forestry, Czech University of Agriculture Prague
165 21 Prague 6-Suchdol, Czech Republic
Telephone: +420-2-24382124
Fax: +420-2-20922252
E-mail: michalkova@lf.czu.cz
Internet: http://www.czu.cz/unesco.htm
            http://terrassa.pln.gov:2080/hydrology/conferences.html

GIS modelling for training in decision making on safe agricultural production in contaminated areas (Novozybkov case study)

Linnik, V.[1], Korobova, E.[1], Kuvylin, A.[1]
van der Perk, M.[2], Burrough, P.[2]

[1]*V.I. Vernadsky Institute of Geochemistry and Analytical Chemistry, Russian Academy of Sciences, Moscow, Russia*
[2]*Utrecht Centre for Environment and Landscape Dynamics, Utrecht University, The Netherlands*

Widespread environmental contamination gave rise to one of the key issues of modern agriculture, which is ecologically safe production. Producing safe food is a particular problem in areas that have been radioactively contaminated after the Chernobyl accident in April 1986. Numerous radio-ecological investigations that took place after the Chernobyl accident proved that milk is the most critical local agricultural product enhancing internal doses of radiation. Milk is an important link in the food chain through 1) environmental (soil-plant) conditions, 2) contamination density, and 3) countermeasures. Milk contamination and transfer factors have been thoroughly studied in a number of international projects.

The present study supported by EC Inco-Copernicus project STRESS has used GIS techniques to verify existing models of radionuclide transfer to milk. Modelling has been performed on the example of the collective farms of the Novozybkov region (Bryansk district) situated within the area contaminated after the Chernobyl accident. A database on Cs-137 concentration in milk measured in the collective farms has been kindly provided by the Research and Technical Center "Protection". Available maps used in our study included: 1/ landscape map (Scale 1:200000); 2/ land use maps (Scale 1:25000); 3/ ^{137}Cs contamination maps (Scale 1:25000). Landscape and land use maps were verified for the completeness and accuracy of the incorporated area and soil information important in radioecological modelling.

For modelling on the district level a grid resolution of 100x100 m was used, for modeling on the farm level, a grid resolution of 50x50 m was used. However the raster size proved to have no significant influence on the accuracy of our cartographic modelling. Comparative analysis of the field and air-gamma survey data performed for settlements of the Novozybkov region revealed the possibilities and limits of air-gamma survey data set in modeling contamination of agricultural products and milk in particular. Computer simulation of milk contamination was based on a deterministic dynamic model using a ^{137}Cs transfer factor for the soil- milk and soil-grass-milk systems. Identification of the parameters of the radioecological model was carried out for several collective farms situated in the Novozybkov district. The accepted model was verified using collective farms with different kinds of soil cover. Prediction of milk contamination was performed for several years in both deterministic and stochastic (for the soil-to-grass link) variants and compared with the radionuclide permissible levels in milk and real concentrations obtained on the farm level. The results of modelling matched the measured values satisfactorily.

This study demonstrated that GIS-based modelling is a convenient tool for agricultural radioecological analysis and prediction. It is believed to be useful in higher agricultural education and real decision making in rural development on the contaminated areas to provide safe agricultural production.

Optimal control of the granulating process of fertiliser in the fluidised bed

Novikov, A .N., Korniyenko, B.Y.

National Technical University of Ukraine, Kiev Polytechnic Institute, Faculty of Physics and Technology, pr. Peremogy 37, 03056 Kiev, Ukraine.

Granulated fertilisers are widely used in agriculture. The production of granulated ammonium sulfatum from the waste products of caprolactum production is the point of special interest. Optimum control of the process with complicated character of phase movement, intensive fluctuations of different types, specialities of drying and crystallisation is an important question. The algorithm of optimum control of the process is based on minimisation of quadric quality criteria.

With an optimum control of the granulating process in the fluidised bed, the process is considered to be two-phased, taking into account basic specialities of real process, which are known in this time. To control the process the mathematical model of granulating in the fluidised bed is used. Therefore three main mechanisms of mixing solid phase in inhomogeneons fluidised bed are known: circulating (connective) vertical movement, turbulent diffusion with a constant speed in persistent phase which is getting down, horizontal change between trains of gas bubbles and other persistent phase.

While the process of drying is running in the first period distribution of mass among granules with their dimensions is calculated, according to the changes in stratum temperature. While the process of drying is running in the second period intensiveness of it is high enough to consider gradient of contents of moisture and changes in contents of moisture in time inside granule equals zero.

As the criteria of quality optimisation has taken discrepancy of the temperature of the fluidised bed with the given one, which maintains stability of granulating process. In order to control influence is chosen an expenditure of given solution. For the solution of the formulated problem variational methods are applicable.

As a result of solution of a posed problem the system of optimal control of process granulating in fluidized bed stratum is developed. There are obtained an optimal control law by the consumption of a solution and optimal distribution of temperatures. The conducted computing experiments confirm functionability of an offered control system.

References
Borodulya V.A., Tepletskey Yu.S., Yepanov A.G., (1981). Particles mixing and carrying a heat in the heterogeneous boiling layers. Minsk, Science and technology,.-41 p.
Todes I.I. (1980) The two-parameters model for mixing a hard phase in fluidised bed. TOXT.. 14 (1), pp.139-144.
Bourovoy E.A., (1969). Automatic processes control in boiling layer. -M.: Metallurgy, 472 p.
Azjogin V.V., Zgurovskiy M.Z., Korbitch Yu.S., (1988). The methods for the stochastic process filtrations and control with distributed parameters: Educational reference - book. - Kiev: "Vyscha Shkola",. 448 p.
Azjogin V.V., Zgurovskiy M.Z.., (1986). Computer aided design of mathematical provision CAD TP. - Kiev: "Vyscha Shkola", 335 p.p

Agricultural education and informational / technological reconstruction of plant growing

Shepovalov, V.

Timiryazev Agricultural Academy in Moscow, Member of International Academy of Agricultura, Education and International Academy of Information

In the beginning of the 21st century agricultural education faces problems of technological progress in the form of radical changes and globalisation of information in the technology of plant growing.

The use of satellite information, the latest achievements in GIS and computerised control over agricultural equipment allows agricultural procedures to be adjusted according to the real conditions of each particular field, taking into consideration the ecological safety of agricultural activities and the stability of local ecosystems.

Precise geographical pin-pointing of agricultural knowledge guarantees wide adaptation and readiness for information interchange which has never been achieved before. All this requires radical changes to the educational programmes and organisation of the process of educating agricultural specialists.

The professional knowledge of future specialists should be narrowed down to particular fields where they will work.

It is necessary to include in the educational programmes the basics of GIS, GPS and methods and technologies for creating electronic maps with agricultural technological conditions.

Future specialist should be professionally and mentally ready to work in conditions of automated protocoling of decisions they make and global monitoring of the ecological effects of industrialised agriculture.

The society of Agricultural Universities should come up with an initiative for developing international requirements to agricultural systems for automated global positioning and standards for representing the geographically oriented agricultural data.

Both businessmen and politicians expect that the sharing of information in the agricultural industry will produce a radical increase in the efficiency of decision making. It is up to us in high agriculture education to respond to this new challenge of technological progress and make it possible for the agricultural industry to meet these expectations.

Adapting Computer-Based Learning Methods in Agricultural Education

Tzortzios S., Adam G.

University of Thessalia, Faculty of Agriculture – Crop and Animal production, Lab of Biometry,Pedion Areos 383 34 Volos, Greece, Tel/Fax: +30 421 74259.

Current advances in information technology have taken educational systems literally to another dimension. In particular, one of the sectors that has benefited from this enormous evolution is agricultural education. A number of products have been developed and applied in agricultural education and research. However, still there are a lot of case problems that need to be tackled further in order to provide some solutions. The proper use of the new Information Technology in the field of agricultural sciences could lead to an educational system based on a method of "Inductive Education" similar to the process of the agricultural research (Tzortzios, 1998), which was presented in "Intagred-99" (Tzortzios and Adam, 1999). In such a process, as the inductive training, the computer has become the most useful tool. First, effectively designed computer programs facilitate the manipulation of ideas and findings, making this process fast and tireless. Second, such programs running in high-speed computers produce an outbreak in statistical ability. Third, these programs have made it possible to test scientific theories based on huge number of variables, which was practically impossible to be handled some years ago. The whole job becomes particularly easier when the various programs, as applications, are constructed in the format of a unique system of conventions by which the user could interact with the programs. A well designed system could allow the user to execute a series of jobs in the least of the time spent for the manual's advice and data handling.

Being in a well-organised database the scientist - teacher or student – could attempt various data manipulations in order to derive interesting specific parameters; to create new variables for various applications; to aggregate groups of certain purposes; and so on. The practical meaning of the users familiarisation with such data handling utilities is obviously of great educational importance, because: (a) it helps the user in developing the necessary self-confidence in approaching the material in study, (b) it offers the chance for a better understanding of the statistical material's meaning as the unique source of any research study; and (c) - the most important – it contributes to the gradual development of the proper scientific mind as a pre-requisite of a possible more integrated form of various complicated problems later on.

After all, going on to the stage of the statistical analysis level of learning, it could be supported that a system based on a more theoretical approach would be rather boring and not far from a negative result regarding applied biological sciences as it is that of agriculture. A system more practical, which suggests a straight involvement to the problem will be much more encouraging and because of this it will be expected to create satisfaction. Some years ago it was rather difficult – if not impossible – to undertake a such approach because of the complete lack of the proper equipment. It is however, incredible nowadays to forget the exploitation suggested, after a such enormous development of the science of computing statistics and its available utilities as important tools in applying biometrical approaches in the field of agriculture.

In this study an attempt was undertaken to approach an integration of various agricultural databases (of plant and animal experiments), multimedia presentations of agricultural data, data analysis tools, etc., using an interactive information system created for this purpose and called *AgroModel*. The aim was to provide a proper educational and friendly environment for efficient agricultural data organisation and manipulation in order to be used for educational and research purposes.

AgroModel was created to provide tools for database management, data manipulation and data analysis, in order to be used initially for educational purposes and later on for research approaches as well. The system was built using object-oriented visual languages (an integration of Visual Basic (Craig C. J. and Webb J., 1998), SQL and Web development languages-HTML, Java) used also for advanced programming (Law A. and Kelton D., 1991) and work with the system. In addition, based on primary agricultural educational and research needs, an appropriate database structure was created on a relational model scheme, where various plant and animal field and experimental data were stored into certain database structures. The database is under continuous improvement and updates since new data and findings are to be continuously added.

The environment provides build-in tools as well as interfaces to software packages (e.g. SPSS, SYSTAT, STATISTICA) for mathematical and statistical analysis applied mainly to agricultural data, but also to any type of data online on the Web. Being in such an integrated environment, where the compatibility between the databases and various statistical packages is available, the trainee could undertake any type of statistical analysis from the simplest descriptive statistics to the most sophisticated statistical approaches (regression models, cluster analysis, etc.). The system comprises a flexible structure that is under continuous improvement and evolution in order to maintain the research information up to date. The overall integrated environment obtained is the platform which is finally to be placed on the WWW in order to provide distance learning education on the web, making the agricultural information available for the benefit of the agricultural society and any scientific society in general.

Arising out of a need for a better organisation and exploitation of agricultural knowledge in higher education, an integrated interactive multimedia environment for agricultural education and research purposes was developed and used efficiently in local area research cases. This integrated system aims to be the model platform used for higher education and research in agriculture and environmental studies in general.

References:
Craig Clark John and Webb Jeff, (1998)."Microsoft Visual Basic 6.0 Programmer's Toolbox". 5th Edition, Microsoft Press, 1998.
Law M. Averill, Kelton W. David, (1991)."Simulation Modelling and Analysis", McGraw-Hill International Editions, Industrial Engineering Series, 1991.
Tzortzios, S. (1998). "Biometry by the use of computers", Teaching Notes. University of Thessalia Press.
Tzortzios S. and Adam G.. (1999) "Proper computer procedures (AgroModel) in plant and animal selection for research and educational purposes", In: *INTAGRED-99 International Workshop Information Technology in Agricultural Education*, 27-29 September 1999, Moscow, Russia.
Valavanis K. and Saridis G..(1990) "A review of intelligent control based methodologies for modelling and analysis of hierarchically intelligent systems", In: *Intelligent Control: Proceedings of the 5th International Symposium*, Northeastern University, MA, 1990, IEEE Computer Society Press.

Farmer use of the internet and the implications for technology-supported distance learning

Warren, M.

Seale-Hayne Faculty of Agriculture, University of Plymouth, Newton Abbot, Devon. TQ12 6NQ. UK

The spatial separation of farm and other small rural businesses has encouraged providers of education and training services to think increasingly in terms of distance learning as a delivery medium. Another incentive (in some cases the most important) has been the opportunity for providers to save on costs of delivery.

The advent of the internet has swelled the ranks of those advocating the distance learning route. Judged in abstract, the internet has the capability to transcend difficulties such as those arising from location of learners, and can allow swift and effective feedback. Increasing capabilities of internet technology allow transmission of multimedia learning packages incorporating music, film and high-resolution images.

There is, however, a danger that the power of such attractions obscures the real difficulties encountered by end users, and that a great deal of useful effort goes to waste through a failure to design systems and information to meet users' needs, rather than the providers' capabilities. This paper draws on an ongoing study at the University of Plymouth, involving longitudinal surveys of farmers in contrasting areas of England, and a case study where farmers were given the facilities to use the email and world-wide web over a period of a year. It also refers to evaluation studies of ICT-based extension and training projects in the South West of England.

Conclusions are drawn concerning the conditions necessary for the increased and effective use of internet technology by farmers, and those conclusions are applied to the specific case of technology-supported learning.

Farmers' Opinions about Programs of Agricultural Experience for Children

Yamada, I.

Rural Life Laboratory, Department of Agricultural Development, National Agriculture Research Center, The Ministry of Agriculture, Forestry and Fisheries of Japan, 3-1-1, Kannondai, Tsukuba, Ibaraki 305-8666. JAPAN
tel & fax: +81-298-38-8419

Recently it has become to be considered educational essential for children to experience agricultural work in some aspects in Japan. For example, children understand food, farming, and ecosystems through farm work. It is also expected to cultivate children's powers of observation, curiosity, etc. Actually such programs go on increasing in various places in Japan. But there are some problems concerning available farmland or agricultural guidance, and they seek for farmers as co-operator. Therefore, the purpose of this study is to clarify farmers' opinions about programs of agricultural experience for children.

1. What do farmers think about effects and advantages of the programs?
2: How many farmers are willing to co-operate?
3: What are conditions to get their agreement?
4: Why they do not want to co-operate?

The data were got from a questionnaire survey of all farming families in the western part of Ueda City in Nagano Prefecture, central part of Japan. The area in the city has both rural and urban character. The survey was conducted in November 1999, and 2247 (83%) were collected.

The attributes of respondents are that: 73 % of respondents are head of a household. Age consist of 2.3% under 39, 15.0% 40-49, 24.6% 50-59, 28.8% 60-69, 22.3% 70-79,and 7.0% over 80. Planting are crop (almost all rice) 70.4%, vegetables 2.6%, flowers 6.2%, fruits 11.2%, others 9.6%. As for the farmers planting rice, 46% of them have less than 20ac rice, in other words most of farming are small scale.

As a result, many farmers think that programs of agricultural experience for children are not only educational but also effective to promote understanding of agriculture. More than half respondents think it is fun for them to come in contact with children, and they will cooperate depending on conditions if they are asked to do. Conditions they think are like these: as assistance, for elementary schoolchild, in their region, and reward is not necessarily. On the other side, farmers who don't want to co-operate said that's because they don't have enough time to do or they are not interested .

Farmer-resident relationship is a key to develop rural area in transition. Programs of agricultural experience for children have a possibility to contribute it.

Concluding Paper

Summary and conclusions

F. Harper

Seale-Hayne Faculty, University of Plymouth, Newton Abbot, Devon TQ12 6NQ, UK

Background and context
This conference was the 5th in the series and the theme arose from the discussions at the Moscow Conference in 1998. There was much debate in Moscow about the shifting focus of the curriculum content, in many institutions and countries, away from the traditional agricultural subjects to greater emphasis on wider issues relating to rural communities and economies. This shift seems to be occurring in countries throughout geographical Europe and in other parts of the world. In Central and Eastern European (CEE) countries a major stimulus has been the political change which led to a market-orientated as opposed to a command economy. In EU countries, the major stimulus has come from the over production of many agricultural commodities at considerable expense to taxpayers. Universities and other centres of Higher Agricultural Education (HAE) must reflect these changes in the curricula which they offer. They must be able to respond rapidly and, where possible, anticipate changes.

The programme
The theme of this Conference was "From Production Agriculture to Rural Development : Challenges for Higher Education in the New Millennium". The plenary sessions had two aims:
- to identify the issues arising from the shift towards rural development; and
- to provide pointers on how centres of HAE should respond to the issues.

The texts of the papers relating to these sessions are included in these proceedings and are self-explanatory.

Six parallel sessions were held in the Conference Programme:
- Agricultural education in transition – facing up to the issues;
- Agricultural education in transition – benefits and experiences from case studies;
- Facing change in the curriculum and its delivery;
- Curriculum changes at the institutional and regional level;
- Agricultural to rural development education in practice; and
- Educational innovations workshop.

Abstracts of the 40 papers presented and discussed in these sessions are published in these proceedings. The Chair of each session was asked to present some general conclusions and themes which emerged. Oral feedback was given in the closing session of the Conference and written comments were submitted later. This feedback has been used in this summary.

What are these conferences about?
At the end of an intensive Conference programme it is worth reminding ourselves why we were there. In general, such Conferences provide the opportunities for members of the HAE academic community and others to come together to listen to keynote speakers and participate in discussions on a topic of mutual interest every two years. The overall aims of these conferences include the following:

- contribution to the global agenda for HAE in the third millennium;
- formulation of strategies for greater international understanding;
- enhancement of co-operation through networking;
- the sharing of experiences and outcomes from the best current national and international projects; and
- the improvement of the overall quality and relevance of the teaching and learning process.

This 5th Conference provided opportunities for these aims to be achieved by over 100 participants from 25 different countries.

The participants

It is interesting to note the countries from which the participants came and to reflect on the reasons. Some 30 participants came from the Central and Eastern European countries of Russia, Ukraine, Poland, Czech Republic, Hungary, Slovakia, Albania and Latvia. The attendance of many from this group can be partly attributed to the financial support provided by the World Bank and FAO. However, many of them attended without sponsorship. Very few attended from southern European countries, i.e. one from Spain, one from Portugal and three from Greece. Without the presence of the participant from FAO, there would have been no-one from Italy. Eight participants came from North America, one from South Africa and one from Australia. The United Kingdom was quite well represented.

It is difficult to attract more than 100 participants from outside the host nation to such Conferences. Whilst the quality of the papers and discussion was high, it may have been even better with double the number. Organisers of future Conferences may wish to consider how to attract more participants.

Times of transition – change

It was evident that major changes are occurring in rural economies and communities throughout geographical Europe. It is still a period of transition and this state is likely to be ongoing. Several papers at this Conference emphasised the changes which are occurring in the rural areas and the HAE curriculum of many different countries. It was also obvious that such changes are more radical, are happening at a greater pace and, arguably, are more necessary in the countries of Central and Eastern Europe.

The pressure for change in education systems is high and many have argued that it is necessary for survival. Change is essential and inevitable. The management of such change is an important issue in HAE and may be associated with resistance and difficulties. Change at a national level has been achieved and the examples of the former Gödöllö University of Agricultural Sciences in Hungary and Wageningen Agricultural University in the Netherlands were cited in the Conference papers. In addition changes are occurring at regional level and within individual universities in many European countries.

There is a danger of changing too quickly and in the wrong direction. It is essential to develop a clear strategy for change before addressing issues of structure. The consequences of change for staff and students must be clearly addressed at an early stage.

There are those who feel that agriculture, and more specifically, agricultural sciences, still have a place in the HAE systems of European countries. That may be so and we must be careful not to neglect these areas. However, there was an overwhelming message from this Conference that change is essential. The change may be towards rural development, as defined by some keynote authors, or it may be into subject areas such as environmental science, food science, business and management or marketing. Individual countries and organisations will determine their own direction and pace of change with varying emphases on different subject areas. It is important not to throw away what is good and still relevant. There is no single definition of rural development which is meaningful enough to totally embrace all of the changes which are going on.

Changes in the curriculum could be achieved in different ways:
- reactive or proactive;
- evolutionary or revolutionary.

Whichever emphasis prevails it is important that the systems of HAE are responsive enough to incorporate change as and when it is necessary. In addition, those in management positions need to know when and how to change. This will involve consultation with various groups, e.g.   employers of graduates;
- government bodies;
- professional associations;
- university staff;
- students; and
- other relevant and specific bodies.

Contributions to the global agenda for HAE

One of the major functions of these Conferences is to identify areas and topics which should be on the global agenda for HAE institutions in Europe and elsewhere and which should feature in the curriculum. This Conference has again yielded a valuable list for this agenda. Some items repeat those from previous Conferences whilst others are new. The more important ones are listed here, but are not in any order of priority or importance.

1. There is a continuing shift towards **inter-disciplinarity** in the curriculum. This broad spectrum of subjects now includes economics, marketing, management, sociology, food science and environmental subjects in what were traditionally straight agriculture courses.

2. There is widespread acceptance that **social** and **political subjects** are now just as important in the curriculum of HAE institutions as the biological, chemical and physical sciences.

3. Whatever the blend of subjects in the curriculum, it must be **relevant to the needs** of students and society in general.

4. It is important for HAE institutions to **know their markets**. This refers to the more traditional markets as well as the need to address the needs of those who can only **learn at a distance** by **part-time mode**. This aspect should also address the requirements of many governments to provide for **lifelong learning** in the population.

5. There was an underlying theme of the need to develop **an entrepreneurial spirit** in those who will be educated and trained to work in rural areas.

6. The topic of **sustainable rural development** arose in several sessions of the Conference and is clearly an important one for inclusion in the curriculum in future. In another context, it was referred to as agricultural and rural sustainability. The discussion

centred on how to achieve this elusive state and how HAE establishments should deal with it. There is clearly a need for more work in this area.

7. Much can be gained from **working closely with partners** in periods of rapid transition. Linking with other centres can result in synergy. There were many examples of **successful partnerships** presented at the Conference. In an extended form, such partnerships lead to **networks** working on areas of common interest, e.g. mountain areas. Such networks can be very productive and **remove some duplication of effort** in a cost-effective way.

8. Rapid change can place pressure and additional demands on university teachers. This is an important area and more effort will need to be put into improving **the competency of teachers** and **their ability to adapt to changes** in the curriculum and its delivery.

9. The workshop on educational innovations demonstrated that there is a continuing move towards more **student-centred learning approaches**. Furthermore, the potential of the **internet and e-based delivery systems** is high particularly for reaching **remote audiences**. Rapid and exciting developments are expected in this area in the next five years.

10. There is a need to **assess the quality of learning and teaching** in HAE institutions. This subject did not receive great attention in this conference. **Quality** of educational provision should **be monitored and assessed** and there is a need for **more feedback** from the recipients and for **self-evaluation**.

Many other and more detailed points emerged during the conference which merit further consideration by a wider audience. The link between **education** and **research** in HAE is important but did not receive great consideration in this Conference.

Co-ordination of the discussions on the future of HAE

It was clear from discussions at the Conference that other groups have been and are actively engaged in discussions on the future needs and direction of HAE.

1. Conferences on Higher Agricultural Education. This is the 5th in the series which started in 1992. Approximately 100 participants attend each conference and the proceedings are published.

2. The inter-university Conference for Agriculture and Related Sciences (ICA). This is a relatively recent group attended by Rectors, Vice-chancellors and other senior members of HAE institutions in Europe.

3. Organisation for Economic Co-operation and Development (OECD). Copies of the Summary Report, including Conference Conclusions and Recommendations, of the Second OECD Conference of Directors and Representatives of Agricultural Knowledge Systems (AKS) held in January 2000 were available to the participants of this Conference. The theme of the Conference was Agricultural Knowledge Systems Addressing Food Safety and Environmental Issues.

4. The World Bank's Rural Development Network, Agricultural Knowledge and Information Systems thematic team carried out an Agricultural Education Review in 1998. Three papers were produced:
- Agricultural Education Review Part I : Past and Present Perspectives
- Agricultural Education Review Part II : Future Perspectives
- Enhancing Agricultural and Rural Education and Training Systems in Rural Development Strategies and Projects.

5. The Food and Agriculture Organisation (FAO) of the United Nations has developed its own policies and statements on future needs for agricultural education, rural development and food security in developing countries.

6. UNESCO, through the Division of Higher Education, organised a World Conference on Higher Education (October 1998) which issued a "World Declaration on Higher Education for the Twenty First Century" and "Framework for Priority Action for Change and Development in Higher Education". Follow-up committees were recommended and it has been suggested that Higher Agricultural Education should be the topic for one of the *ad hoc* subject groups.

7. International Association of Agricultural Students (IAAS). This Association provides a forum for agricultural students around the world to develop their own views and policies. It was very helpful to have the new President of the IAAS at this Conference to give a brief presentation. It is vital that those who initiate and implement change seek the views of students and listen to them.

There is no shortage of groups and associations which have discussed and written reports on the future of Higher Agricultural Education. I know of no organisation which co-ordinates the outcomes of these seven, and possibly other, groups. There is a need for such a body in order to disseminate all of the information to all of the interested parties and to avoid duplication of effort. The International Organising Committee should address this issue, formulate its views and state its position.

In conclusion

This summary report represents a personal view on this Conference at the University of Plymouth. Professor Denis Lucey, University College Cork, Eire, was scheduled to give a summary presentation at the Conference and write this report. He was unable to attend because of pressing personal reasons. He continued to provide input to the Conference from a distance and arranged for the copies of the OECD Summary Report referred to in 3 above to be delivered to the participants. I am grateful to Professor Lucey for his input in what was a very difficult time for him. The participants ensured the success of the Conference through their high quality and enthusiastic contributions. I hope that many of the points raised in this summary will appear on the global agenda for HAE and help to steer it through the rapidly changing 'environment' of the early years of the twenty-first century.

Many good things are happening in HAE institutions and no doubt some mistakes have been made. One of the great benefits of these conferences is that participants can share experiences and learn from each other. I believe that this 5th Conference provided such an opportunity and also created a better international understanding through cultural and social activities.

There will be much to talk about and enjoy in Greece, 2002.

List of Participants

arranged alphabetically by country

ECHAE5 Participants list by country.

Name and e mail	Address	Country
Dr Ylli **BICOKU** ylli@ifdc.albnet.net	International Fertiliser Development Rruga "Mihal Duri" Nr. 17/5 Tirana	Albania
Dr Alan **WEARING** alan.wearing@mailbox.uq.edu.au	University of Queensland Gatton Queensland 4343	Australia
Dr Oliver **MEIXNER** meixner@edv1.boku.ac.at	Institut fur Agraroekonomik Universitat fur Bodenkultur Wien Peter Jordan-Str.82 Vienna A-1190	Austria
Prof Dr Adolf **ZAUSSINGER** zaussinger@mail.boku.ac.at	Dean of Studies University of Agricultural Sciences Nussdorfer Laende 29-31 Vienna A-1190	Austria
Ms Maria **ERLANDSSON** a5marerl@ulmo.stud.slu.se	IAAS (Sweden) Kardinaal Mercierlaan 92 Heverlee B-3001	Belgium
Prof Ir Guido **VAN HUYLENBROECK** guido.vanhuylenbroeck@rug.ac.be	Department of Agricultural Economics Ghent University Coupure Links 653 Gent 9000	Belgium
Ms Ann **COONEY** acooney@agr.gov.sk.ca	College of Agriculture Department of Agricultural Economics University of Saskatchewan Saskatoon Saskatchewan S7N 5A8	Canada
Mr Robert **CUMMING** cumminb@em.agr.ca	Room 4112 Agriculture & Agri-Food Canada 930 Carling Avenue Sir John Carling Building Ontario Ottawa KIA 0C5	Canada
Prof Kevin **PARTON** kparton@agec.uoguelph.ca	Ontario Agricultural College University of Guelph Department of Agricultural Economics Guelph Ontario N1G 2W1	Canada

ECHAE5 Participants list by country.

Name and e mail	Address	Country
Prof Peter **STONEHOUSE** stonehou@agec.uoguelph.ca	Department of Agricultural Economics & Business University of Guelph 64 Vanier Drive Guelph Ontario N1G 2W1	Canada
Prof Pavel **KOVAR** kovar@les.czu.cz	Department of Land Use and Improvement Czech University of Agriculture in Prague Forestry Faculty Suchdol Prague -6 165 21	Czech Republic
Professor Josef **KOZAK** vlkova@lf.czu.cz	Rector's Office Czech University of Agriculture in Prague Suchdol Prague-6 165 21	Czech Republic
Ass Prof Michal **LOSTAK** lostak@pef.czu.cz	Department of Humanities Czech University of Agriculture in Prague Kamycka St. 12, 129 Suchdol Prague-6 165 21	Czech Republic
Dr Milan **SLAVIK** slavik@chuchle.czu.cz	Department of Education Czech University of Agriculture in Prague V Laznich-3 Mala Chuchle Prague-5 159 00	Czech Republic
Dr Brian **DENNIS** bde@kvl.dk	Department of Agricultural Sciences Royal Veterinary and Agricultural Agrojev 10 Taastrup DK-2630	Denmark
Prof Flemming **FRANDSEN** kvl@fnf.dk	Ecology and Molecular Biology Royal Veterinary and Agricultural Bulowsvej 17 Frederiksberg C Copenhagen DK-1870	Denmark
Mrs Minna **MIKKOLA** minna.mikkola@helsinki.fi	Faculty of Agriculture and Forestry University of Helsinki PB 62 Helsinki 00014	Finland

ECHAE5 Participants list by country.

Name and e mail	Address	Country
Prof Bruno **GUERMONPREZ** bguermonprez@isa.fopl.asso.fr	Institut Superieure d'Agriculture (ISA) 13 rue de Toul Lille Cedex 59146	France
Dr Christian **SCHVARTZ** christian.schvartz@isa.fupl.asso.fr	Farming System Department Institut Superieure d'Agriculture (ISA) 41, rue du Port Lille Cedex 59146	France
Mr Artis **KANCS** kancs@iamo.uni-halle.de	Institut fur Agrarentwicklung Theodor-Lieser-Str. 2, Halle D-06120	Germany
Dr Volker **MOTHES** mothes@iamo.uni-halle.de	Institute of Agricultural Development Theodor-Lieser-Str. 2, Halle D-06120	Germany
Prof Evangelos **KAPETANAKIS** ekapet@lyttos.admin.teiher.gr	Technical Education Institute Stavromenos P.O. Box 140 Heraklion Crete GR-71110	Greece
Dr Dimitrios **KOLLAROS** ekapet@lyttos.admin.teiher.gr	Technical Education Institute Stavromenos P.O. Box 140 Heraklion Crete GR-71110	Greece
Dr Alex **KOUTSOURIS** alex@kar.forthnet.gr	Department of Rural Economics and Development Agricultural University of Athens 75 Iera Odos Athens 11855	Greece
Ms Katalin **BALAZS** balazsk@nt.ktg.gau.hu	Institute of Environmental Management Szent Istvan University Pater K.U.1 Godollo H-2103	Hungary
Dr Gyorgy **FULEKY** fuleky@fau.gau.hu	Szent Istvan University Pater K.U.1 Godollo H-2103	Hungary

ECHAE5 Participants list by country.

Name and e mail	Address	Country
Prof Csaba ILLES illes@GTK-F1.gau.hu	Szent Istvan University Pater K.U.1 Godollo H-2103	Hungary
Dr Jozsef KISS jkiss@gikk.gau.hu	Szent Istvan University Pater K.U.1, Godollo H-2103	Hungary
Ms Izumi YAMADA iyamada@narc.affrc.go.jp	Rural Life Laboratory National Agriculture Research Center 3-1-1 Kannondai Tsukuba Ibaraki 305-8666	Japan
Prof Dr Peteris BUSMANIS	Latvia University of Agriculture 2 Liela iela Jelgava LV-3001	Latvia
Prof Baiba RIVZA aip@latnet.lv	The Council of Higher Education of Latvia 21 Meistaru St. Riga LV-1050	Latvia
Dr Jan DRIESSE j.driesse@pers.vhall.nl	van Hall Institute AGORA Postbus 1753 Leeuwarden NL-8901 CB	Netherlands
Mr Teo DUNNING t.dunning@iahldev.agro.nl	Larenstein International Agricultural P O Box 7 Deventer NL-7400 AA	Netherlands
Rector Cees KARSSEN cees.karssen@rvb.kcw.wau.nl	Central Office Wageningen University P O Box 9101 Wageningen NL-6700 HB	Netherlands
Dr Wiebe NIJLUNSING w.nijlunsing@pers.vhall.nl	Business Centre van Hall Institute Postbus 1754 Leeuwarden NL-8901 CB	Netherlands

ECHAE5 Participants list by country.

Name and e mail	Address	Country
Prof Stanislaw DRZYMALA drzymala@au.poznan.pl	Faculty of Agronomy Agricultural University of Poznan ul. Wojska Polskiego 28 Poznan PL-60-637	Poland
Dr Szczepan FIGIEL sfigiel@icbpm.uwm.edu.pl	International Center for Business and Public Management Warmia and Masuria University in Olsztyn ul. Prawochenskiego 19, Olsztyn PL-10-720	Poland
Prof Antoni JOWKO jowko@ap.siedlce.pl	Agricultural and Teachers University of ul. 3 Maja 54 Siedlce PL-08-110	Poland
Dr Stanislaw PILARSKI spilarski@icbpm.uwm.edu.pl	International Center for Business and Public Management Warmia and Masuria University in Olsztyn ul. Prawochenskiego 19, Olsztyn PL-10-720	Poland
Dr Andrzej PILICHOWSKI pilan@krysia.uni.lodz.pl	Institute of Sociology University of Lodz 41 Rewolucji 1905 r str Lodz PL-90-214	Poland
Dr Janusz ROWINSKI wigierm@ierigz.waw.pl	Polish Academy of Sciences Pilicka 23 m.1 Warsaw PL-02-613	Poland
Dean Grzrgorz SKRZYPCZAK gsweed@au.poznan.pl	Faculty of Agronomy Agricultural University of Poznan ul. Wojska Polskiego 28 Poznan PL-60-637	Poland
Dr Zbigniew WARZOCHA zwarzocha@icbpm.uwm.edu.pl	International Center for Business and Public Management Warmia and Masuria University in Olsztyn ul. Prawochenskiego 19, Olsztyn PL-10-720	Poland
Prof Dr Tadeusz WIECZOREK	Warsaw Agricultural University Nowoursynowska 166 Warsaw PL-02-787	Poland

ECHAE5 Participants list by country.

Name and e mail	Address	Country
Prof Pedro **LYNCE DE FARIA**	Departmento Produao Agricola e Animal-Agricultura Instituti Superior de Agranomia Tapada da Ajuda Lisbon 1349-017	Portugal
Dr Tom **BRYSON** afta@online.ru sart@online.ru	TACIS Team Leader Timiryazev Agricultural Academy in ul.Timiryazevskaya, 49 Moscow 127550	Russia
Dr Valeri **CHUMAKOV** v.chumakov@g23.relcom.ru	Department of International Programmes Moscow State Agroenginerring University Timiryazevskaya 58 Moscow 127550	Russia
Dr Alexey **IVANOV**	Timiryazev Agricultural Academy in ul.Timiryazevskaya, 49 Moscow 127550	Russia
Prof Igor **KOMISSAROV** agro@sibtel.ru	Tyumen State Agricultural Academy 7 Republic Str. Tyumen 625003	Russia
Prof Dr Tatyana **KORSUNOVA** bgsha@buryatia.ru	Agricultural Ecology Department Buryat State Agricultural Academy Babushkina 32-17 Ulan-Ude Buryatia 670005	Russia
Prof Eugeny **KOSHKIN** sart@online.ru	c/o Timiryazev Agricultural Academy Ministry of Agriculture and Food Timiryazev str. 49 Moscow 127550	Russia
Dr Vassili **LAVROVSKI** v-lavr@aha.ru	Timiryazev Agricultural Academy Timiryazev str. 49 Moscow 127550	Russia
Dr Olivier **NORMAND** afta@online.ru sart@online.ru	TACIS Team Timiryazev Agricultural Academy in ul.Timiryazevskaya, 49 Moscow 127550	Russia

ECHAE5 Participants list by country.

Name and e mail	Address	Country
Ass Prof Lubica **BARTOVA** bartova@uniag.sk	Department of Statistics and Operations Research Slovak Agricultural University in Nitra tr. A. Hlinku 2 Nitra 94976	Slovakia
Dr Maggi **LININGTON** LNGTN-MJ@caleph.vista.ac.za	Department Agricultural Sciences Vista University VUDEC P/Bag X641 Pretoria 0001	South Africa
Dr Eduardo **RAMOS** eduardo.ramos@uco.es	Departmento Economia Agraria. ETSIAM Equipo Desarollo Rural P.O. Box 3048 Cordoba 14080	Spain
Prof Anna **MARTENSSON** Anna.Martensson@ekon.slu.se	Core Curriculum in Agricultural Sciences Swedish University of Agricultural Box 7013 Uppsala SE-750 07	Sweden
Prof Dr Petro **LAKYDA** inter@nauu.kiev.ua	Forest Management Department National Agricultural University of Ukraine 03041, Kyiv, Geroyiv Oborony St., 15 Kiev	Ukraine
Mr Paul **BRASSLEY** pbrassley@plymouth.ac.uk	Seale-Hayne Faculty University of Plymouth Newton Abbot Devon TQ12 6NQ	United Kingdom
Mrs Clare **BROOM** cbroom@plymouth.ac.uk	Dean, Seale-Hayne Faculty University of Plymouth Newton Abbot Devon TQ12 6NQ	United Kingdom
Professor John **BULL** jbull@plymouth.ac.uk	Vice Chancellor University of Plymouth Drake Circus Plymouth Devon PL4 8AA	United Kingdom
Mrs Val **CREAN** vcrean@plymouth.ac.uk	Seale -Hayne Faculty University of Plymouth Newton Abbot Devon TQ12 6NQ	United Kingdom

ECHAE5 Participants list by country.

Name and e mail	Address	Country
Mr Matthew DAVID m.david@plymouth.ac.uk	Department of Sociology University of Plymouth Drake Circus Plymouth Devon PL4 8AA	United Kingdom
Miss Mary EAMES mary-eames@countryside-alliance.org	Education Officer Countryside Alliance The Old Town Hall 367 Kennington Road London SE11 4PT	United Kingdom
Dr John EDDISON jeddison@plymouth.ac.uk	Seale-Hayne Faculty University of Plymouth Newton Abbot Devon TQ12 6NQ	United Kingdom
Prof Andrew ERRINGTON aerrington@plymouth.ac.uk	Seale-Hayne Faculty University of Plymouth Newton Abbot Devon TQ12 6NQ	United Kingdom
Mr Tim FELTON tfelton@plymouth.ac.uk	Seale-Hayne Faculty University of Plymouth Newton Abbot Devon TQ12 6NQ	United Kingdom
Mr Steve FISHER sjfisher@plymouth.ac.uk	Seale-Hayne Faculty University of Plymouth Newton Abbot Devon TQ12 6NQ	United Kingdom
Dr Mick FULLER mfuller@plymouth.ac.uk	Seale-Hayne Faculty University of Plymouth Newton Abbot Devon TQ12 6NQ	United Kingdom
Dr David GIBBON davidpgibbon@freeuk.com	The Spinney Trow Hill Drive Trow Hill Sidford Sidmouth Devon EX10 0PW	United Kingdom
Prof Fred HARPER	Seale-Hayne Faculty University of Plymouth Newton Abbot Devon TQ12 6NQ	United Kingdom

ECHAE5 Participants list by country.

Name and e mail	Address	Country
Dr Simon **HEATH** clues@abdn.ac.uk	Centre for CBL in Land Use & Environmental Sciences University of Aberdeen MacRobert Building Aberdeen Scotland AB24 5UA	United Kingdom
Mr Duncan **HICKS** enquire@hartpury.ac.uk	Hartpury College c/o Principal's Office Hartpury Gloucester GL19 8BE	United Kingdom
Mr Guy **HILLS SPEDDING**	Countryside Alliance The Old Town Hall 367 Kennington Road London SE11 4PT	United Kingdom
Mr Peter **HOLGATE** pholgate@plymouth.ac.uk	Seale-Hayne Faculty University of Plymouth Newton Abbot Devon TQ12 6NQ	United Kingdom
Mr Ian **HUTCHCROFT** ihutchr@devon-cc.gov.uk	Chief Executive's Directorate Devon County Council County Hall Exeter Devon Ex2 4QD	United Kingdom
Dr William **HUTCHEON** b.hutcheon@ab.sac.ac.uk	Scottish Agricultural College Ferguson Building Craibstone Estate Bucksburn Aberdeen Scotland AB21 9YA	United Kingdom
Dr Anita **JELLINGS** ajellings@plymouth.ac.uk	Seale-Hayne Faculty University of Plymouth Newton Abbot Devon TQ12 6NQ	United Kingdom
Prof Philip **LOWE** e.m.curry@ncl.ac.uk cre@ncl.ac.uk	Centre for Rural Economy University of Newcastle-upon-Tyne Newcastle-upon-Tyne Northumbria NE1 7RU	United Kingdom
Mr Nick **MILLARD** nick.millard@bkonline.co.uk	Bruton Knowles Tauntfield 109 South Road Taunton Somerset TA1 3ND	United Kingdom

ECHAE5 Participants list by country.

Name and e mail	Address	Country
Ms Jackie **PALMER** jpalmer@plym.ac.uk	Seale-Hayne Faculty University of Plymouth Newton Abbot Devon TQ12 6NQ	United Kingdom
Dr Julian **PARK** j.r.park@reading.ac.uk	Department of Agriculture University of Reading Earley Gate P O Box 236 Reading Berkshire RG6 6AT	United Kingdom
Mr Steve **PARSONS**	School of Management Harper Adams Agricultural College Newport Shropshire TF10 8NB	United Kingdom
Dr Tahir **REHMAN** t.u.rehman@reading.ac.uk	Department of Agriculture University of Reading Earley Gate P O Box 236 Reading Berkshire RG6 6AT	United Kingdom
John **RUSSELL** jfarussell@compuserve.com	Stoney Reach 36a Countess Wear Road Exeter Devon EX5 1AG	United Kingdom
Dr Barbara **SHEAVES** bsheaves@plymouth.ac.uk	Seale-Hayne Faculty University of Plymouth Newton Abbot Devon TQ12 6NQ	United Kingdom
Mr Richard **SOFFE** rsoffe@plymouth.ac.uk	Seale-Hayne Faculty University of Plymouth Newton Abbot Devon TQ12 6NQ	United Kingdom
Mr Mark **STONE** m2stone@plymouth.ac.uk	Seale-Hayne Faculty University of Plymouth Newton Abbot Devon TQ12 6NQ	United Kingdom
Mr Martin **TURNER**	Agricultural Economics Unit, Lafrowda Exeter University St Germans Road Exeter Devon EX4 6TL	United Kingdom

ECHAE5 Participants list by country.

Name and e mail	Address	Country
Miss Angels **VAREA** mav20@cam.ac.uk	Department Veterinary Medicine University of Cambridge Cambridge Cambridgeshire CB3 0ES	United Kingdom
Mr Martyn **WARREN** mwarren@plymouth.ac.uk	Seale-Hayne Faculty University of Plymouth Newton Abbot Devon TQ12 6NQ	United Kingdom
Dr Jill **WHITE** jillmwhite@compuserve.com	Woodland Cottage Avonwick South Brent Devon TQ10 9ES	United Kingdom
Mr Ian **WHITEHEAD** iwhitehead@plymouth.ac.uk	Seale-Hayne Faculty University of Plymouth Newton Abbot Devon TQ12 6NQ	United Kingdom
Mr Robert **WILLIAMS** rjwilliams@plymouth.ac.uk	Seale-Hayne Faculty University of Plymouth Newton Abbot Devon TQ12 6NQ	United Kingdom
Dr Eirene **WILLIAMS** ewilliams@plymouth.ac.uk ECHAE5@plymouth.ac.uk	Seale-Hayne Faculty University of Plymouth Newton Abbot Devon TQ12 6NQ	United Kingdom
Mr David **WYLLIE** D.Wyllie@sac.ed.ac.uk	Scottish Agricultural College West Mains Road Edinburgh Scotland EH9 3JG	United Kingdom
Dr Thomas **BRUENING** tbruening@PSU.EDU	Pennsylvania State University 335 Ag Admin Bld University Park Pennsylvania PA 16802	United States of America
Prof Drew **HYMAN** dwh@psu.edu	Department of Agricultural Economics and Rural Sociology Pennsylvania State University 6 Armsby Building University Park Pennsylvania 16802 PA	United States of America
Dr Jim **POLSON** polson.1@osu.edu	Ohio State University 1680 Madison Avenue Wooster Ohio 44691	United States of America

ECHAE5 Participants list by country.

Name and e mail	Address	Country
Ms Lavinia **GASPERINI** lavinia.gasperini@fao.org	Senior Education Officer FAO , SDR Division Viale delle terme di Caracalla Roma Italy 00100	**World Organisation**
Dr Charles **MAGUIRE** Cmaguire@worldbank.org	Rural Development Department The World Bank 1818 H Street N.W. Room 137 Washington DC USA 20433	**World Organisation**